PREPARATION G

ASE AUTOMOTIVE PARTS SPECIALIST TEST (P2)

We encourage professionalism through voluntary certification.

Delmar Publisher's Online Services

To access a wide variety of Delmar products and services on the World Wide Web, point your browser to:
http://www.delmar.com (or email: info@delmar.com)

To access International Thomson Publishing's home site for information on more than 34 publishers and 20,000 products, browse:
http://www.thomson.com (or email: findit@kiosk.thomson.com)

PREPARATION GUIDE FOR THE

ASE AUTOMOTIVE PARTS SPECIALIST TEST (P2)

Texas State Technical College
EDIT Department
Technical Writer, Norris Martin
Editor, Robert W. Gentry, Ph.D.

We encourage professionalism through voluntary certification.

Delmar Publishers

an International Thomson Publishing company

Albany • Bonn • Boston • Cincinnati • Detroit • London • Madrid
Melbourne • Mexico City • New York • Pacific Grove • Paris • San Francisco
Singapore • Tokyo • Toronto • Washington

NOTICE TO THE READER

Publisher does not warrant or guarantee any of the products described herein or perform any independent analysis in connection with any of the product information contained herein. Publisher does not assume, and expressly disclaims, any obligation to obtain and include information other than that provided to it by the manufacturer.

The reader is expressly warned to consider and adopt all safety precautions that might be indicated by the activities herein and to avoid all potential hazards. By following the instructions contained herein, the reader willingly assumes all risks in connection with such instructions.

The publisher makes no representation or warranties of any kind, including but not limited to, the warranties of fitness for particular purpose or merchantability, nor are any such representations implied with respect to the material set forth herein, and the publisher takes no responsibility with respect to such material. The publisher shall not be liable for any special, consequential, or exemplary damages resulting, in whole or part, from the readers' use of, or reliance upon, this material.

Delmar Staff
Executive Director: Dale Bennie
New Business Development Manager: Mark Huth
Editor: Jack Erjavec
Production Manager: Dianne Jensis
Marketing Manager: Kathryn Little

COPYRIGHT © 1996
By Delmar Publishers
an International Thomson Publishing Company

The ITP logo is a trademark under license.

Printed in the United States of America

For more information, contact:

Delmar Publishers
3 Circle, Box 15015
Albany, New York 12212-5015

International Thomson Publishing Europe
Berkshire House
168-173 High Holborn
London, WC1V 7AA
England

Thomas Nelson Australia
102 Dodds Street
South Melbourne, 3205
Victoria, Australia

Nelson Canada
1120 Birchmount Road
Scarborough, Ontario
Canada, M1K 5G4

International Thomson Editores
Campos Eliseos 385, Piso 7
Col Polanco
11560 Mexico D F Mexico

International Thomson Publishing GmbH
Konigswinterer Strasse 418
53227 Bonn
Germany

International Thomson Publishing Asia
221 Henderson Road
#05-10 Henderson Building
Singapore 0315

International Thomson Publishing—Japan
Hirakawacho Kyowa Building, 3F
2-2-1 Hirakawacho
Chiyoda-ku, Tokyo 102
Japan

All rights reserved. No part of this work covered by the copyright hereon may be reproduced or used in any form or by any means—graphic, electronic, or mechanical, including photocopying, recording, taping, or information storage and retrieval systems—without the written permission of the publisher.

Portions of materials contained herein have been reprinted with permission of General Motors Corporation, Service Technology Group.

2 3 4 5 6 7 8 9 10 XXX 01 00 99 98 97

Library of Congress Cataloging-in-Publication Data
Preparation guide for the ASE automotive parts specialist test (P2)/Texas State Technical College, EDIT Department; technical writer, Norris Martin; editor, Robert W. Gentry.
    p.   cm.
    ISBN 0-8273-7552-2
1. Automobiles—Parts—Examinations, questions, etc. 2. National Institute for Automotive Service Excellence—Examinations—Study guides. 3. Automobile Mechanics—Certification—United States. I. Martin, Norris. II. Gentry, Robert W., Ph.D. III. Texas State Technical College (Waco, Texas). EDIT Dept.
629.28'7'076—dc20
    95-40817
    CIP

TABLE OF CONTENTS

PREFACE — VII

CHAPTER 1 – INTRODUCTION — 1

    Automotive Parts Specialist Test (P2) Task List — 1

    How to Pass the Test — 6

CHAPTER 2 – SAMPLE QUESTIONS — 9

CHAPTER 3 – ANSWERS AND ANALYSIS — 51

GLOSSARY — 85

APPENDIX — 97

PREFACE

There is little doubt that the Automotive Parts Specialist Test (P2) from ASE is a difficult test. It's difficult because of the wide variety of subjects covered and the technical areas that you must study. This book is designed to make the test less difficult. Every item of ASE's task list for this test is represented by at least one question in this book. Each question has a detailed answer, which not only gives you the correct answer, but also explains why the answer is correct and why the others are wrong.

The ASE P2 Test covers the Parts Specialist's entire job: general operations, customer relations and sales skills, vehicle systems knowledge, vehicle identification, cataloging skills, inventory management, and merchandising. Your knowledge must be broad enough to answer questions in each of these areas.

This study guide is written for anyone who wishes to pass the ASE P2 Test, whether that person has had previous experience or not. The author assumes you have some knowledge of mechanics, but if you do not, you can acquire some of this knowledge while studying this guide.

This study guide is divided into four parts: three chapters and a glossary of terms you should know. Chapter One lists the 92 tasks from which the ASE P2 Test questions will be taken. Every item of the task list is represented by at least one question in this book. Chapter One also outlines some proven methods to help you prepare for and pass the test.

Chapters Two and Three contain the heart of this guide's unique approach—sample test questions and an analysis of the answers. First, Chapter Two presents sample questions for every task covered on the test. There are many more questions in Chapter Two than will appear on the ASE P2 Test. The actual ASE P2 Test is highly confidential, but these questions come as close to the real test questions as possible.

After you have answered the sample questions in Chapter Two, you can see how well you did by reviewing Chapter Three. Chapter Three lists the correct answer for each question, along with an explanation of the question and all of the answers. We highly recommend that you read through the explanation for each question with which you felt uncomfortable. Of course, it wouldn't hurt to read through all of the answers.

Take the sample test some time before you are scheduled to take the real test. Look through the answers and analysis of the questions. Work on all of your weak areas. Then, two or three days before the test, take the sample test again. The results of the test will show you the areas in which you are still weak. Study only those areas before the test. Don't go over the areas you are comfortable with; this extra study may only confuse you.

Good luck!

ACKNOWLEDGMENTS

The author and editor extend a special thanks to the Texas State Technical College Automotive Department faculty for their generous assistance with this text. In addition, thanks are expressed to General Motors Corporation (Service Technology Group), to Pep Boys Automotive Supercenters, and to Automatic Data Processing (ADP) for their important contributions.

CHAPTER 1 – INTRODUCTION

AUTOMOTIVE PARTS SPECIALIST TEST (P2) TASK LIST

Specifications for Automotive Parts Specialist Test (P2)*

Content Area	Questions in Test	Percentage of Test
A. General Operations	11	15.7%
B. Customer Relations and Sales Skills	14	20.0%
C. Vehicle Systems Knowledge	30	42.9%
1. Engine Mechanical Parts	2	
2. Cooling Systems	2	
3. Fuel Systems	3	
4. Ignition Systems	3	
5. Exhaust Systems	1	
6. Emissions Control System	4	
7. Manual Transmission/Transaxle	1	
8. Automatic Transmission/Transaxle	1	
9. Drivetrain Components	2	
10. Brakes	3	
11. Suspension and Steering	2	
12. Heating and Air Conditioning	2	
13. Electrical Systems	2	
14. Miscellaneous	2	
D. Vehicle Identification	2	2.9%
E. Cataloging Skills	5	7.1%
F. Inventory Management	5	7.1%
G. Merchandising	3	4.3%
Total	70	100.0%

*Note: There could be up to ten additional pre-test questions. Your answers to these pre-test questions will not affect your score, but since you do not know which they are, you should answer all questions in the test. The five-year Recertification Test will cover the same content areas as those listed above. However, the number of questions in each content area of the Recertification Test will be reduced by about one-half.

Automotive Parts Specialist Test Task List

A. **General Operations (11 questions)**

Task 1   Calculate discounts/percentages.

Task 2   Calculate special handling charges.

Task 3   Identify and convert units of measure.

Task 4   Determine alphanumeric sequences.

Task 5   Determine sizes with precision measuring tools and equipment.

Task 6   Perform money transactions (cash, checks, and credit cards).

Task 7   Perform sales and credit invoicing.

Task 8   Interact with management and fellow employees.

Task 9   Demonstrate housekeeping skills (facility, work stations, and backroom).

Task 10 Assist with new employee training.

Task 11 Demonstrate proper safety practices.

Task 12 Identify proper handling of regulated and/or hazardous materials.

Task 13 Identify potential security risks.

Task 14 Identify parts industry terminology.

B. **Customer Relations and Sales Skills (14 questions)**

Task 1   Identify customer types (do-it-yourself and professional installer).

Task 2   Identify customer needs.

Task 3   Provide technical and other information.

Task 4   Handle customer complaints and returns.

Task 5   Acknowledge customer.

Task 6   Demonstrate proper telephone skills.

Task 7   Obtain pertinent application information.

Task 8   Present professional image.

Task 9   Recommend related items.

Task 10   Identify product features and benefits.

Task 11   Handle objections.

Task 12   Balance telephone and in-store customers.

Task 13   Promote store services and features.

Task 14   Promote upgraded products.

Task 15   Solve customer problems.

Task 16   Close sale.

C. **Vehicle Systems Knowledge (30 questions)**

   1. **Engine Mechanical Parts (2 questions)**

      Task 1   Identify major components.

      Task 2   Identify component function.

   2. **Cooling Systems (2 questions)**

      Task 1   Identify major components.

      Task 2   Identify component function.

   3. **Fuel Systems (3 questions)**

      Task 1   Identify major components.

      Task 2   Identify component function.

   4. **Ignition Systems (3 questions)**

      Task 1   Identify major components.

      Task 2   Identify component function.

   5. **Exhaust Systems (1 question)**

      Task 1   Identify major components.

      Task 2   Identify component function.

   6. **Emissions Control System (4 questions)**

      Task 1   Identify major components.

      Task 2   Identify component function.

   7. **Manual Transmission/Transaxle (1 question)**

      Task 1   Identify major components.

      Task 2   Identify component function.

8. **Automatic Transmission/Transaxle (1 question)**

    Task 1   Identify major components.

    Task 2   Identify component function.

9. **Drive Train Components (Includes driveshafts, halfshafts, U-joints, CV joints, and four-wheel drive systems) (2 questions)**

    Task 1   Identify major components.

    Task 2   Identify component function.

10. **Brakes (3 questions)**

    Task 1   Identify major components.

    Task 2   Identify component function.

11. **Suspension and Steering (2 questions)**

    Task 1   Identify major components.

    Task 2   Identify component function.

12. **Heating and Air Conditioning (2 questions)**

    Task 1   Identify major components.

    Task 2   Identify component function.

13. **Electrical Systems (2 questions)**

    Task 1   Identify major components.

    Task 2   Identify component function.

14. **Miscellaneous (2 questions)**

    Task 1   Identify fastener thread types (SAE, USS, and metric).

    Task 2   Identify fastener thread diameter and pitch.

    Task 3   Identify fastener type.

    Task 4   Identify fastener grade.

    Task 5   Identify fitting type.

    Task 6   Identify fitting sizes.

    Task 7   Identify body repair and refinishing materials and supplies.

    Task 8   Identify hose and tubing types and applications.

    Task 9   Determine hose and tubing size.

    Task 10   Recommend proper application and usage of chemicals.

D. Vehicle Identification (2 questions)

Task 1   Locate Vehicle Identification Number (VIN).

Task 2   Locate production date.

Task 3   Locate component identification data.

Task 4   Identify body styles.

Task 5   Utilize additional references.

Task 6   Locate paint code(s).

E. Cataloging Skills (5 questions)

Task 1   Locate proper catalog.

Task 2   Obtain and interpret additional information.

Task 3   Utilize additional reference material (technical bulletins, interchange list, supplements, etc.).

Task 4   Identify catalog terminology and abbreviations.

Task 5   Locate index and table of contents.

Task 6   Perform catalog maintenance.

F. Inventory Management (5 questions)

Task 1   Report lost sales.

Task 2   Verify incoming and outgoing merchandise.

Task 3   Perform physical inventory.

Task 4   Report inventory discrepancies.

Task 5   Perform stock rotation.

Task 6   Handle special orders.

Task 7   Perform proper core handling (i.e., accepting or declining cores, storage, and return).

Task 8   Handle warranty returns.

Task 9   Determine proper selling unit (each, pair, case, etc.).

Task 10 Handle return of broken kits, special order parts, and exchange parts.

G. **Merchandising (3 questions)**

Task 1   Locate and arrange displays.

Task 2   Display pricing.

Task 3   Inspect and maintain shelf quantities/conditions.

Task 4   Identify impulse, seasonal, and related items.

HOW TO PASS THE TEST

The ASE Automotive Parts Specialist Test (P2) is like all other multiple-choice tests you have taken and that you will take—you need a plan for (1) preparing for the test and (2) passing the test. Keep thinking *"How* will I remember this?", not "I *will* remember this." You already plan to pass the test, not just take it, or you wouldn't have bought this book and set your sights on the test.

This book is constructed like the test so that you can study for the test and practice taking it at the same time. The writers of the sample questions in this book used their experience, their knowledge of several other ASE tests, and their study of the Parts Specialist field to construct questions that are as close as possible to the actual test questions.

What are the advantages of studying sample questions rather than reading a book about the Parts Specialist position? First, answering sample questions tells you what areas you already know and, therefore, don't need to study. On the other side, sample questions show you what you **don't** know, making your study plan simple. Also, sample questions give you an idea of typical wrong answers (called "distractors") that may be on the real test. Since all the possible answers are discussed in Chapter Three, you will become familiar with the answer choices that could mislead you.

The experts all agree that passing any test requires some strategy—consider these suggestions:

Prepare for the test by following several steps:

1. Find a private place to study that is free from distractions such as TV, hobbies, and people. Maybe it's just a corner or an out-of-the-way room. Find the best light and a comfortable (not too comfortable) chair.

2. Take breaks—set time limits for each study session. Every twenty minutes or so, get up and stretch. Walk around a little, but don't eat a big snack—you may get sleepy.

3. Give yourself adequate preparation time. Start several days before the test, and study some each day. Study in bites, not in huge gulps—twenty minutes every day is much better than one long session on Saturday. A

short session each day prepares you for the test, and the key secret to keeping your cool during the test is being prepared. Preparation makes you feel confident.

4. Become familiar with the types of questions on the test. This book shows you all the types you will find. Notice the type of question which begins, "Parts Specialist A says...." This type of question will always be like two true-false questions rolled into one. Look at each half carefully—it's not a tricky question, but it does require careful reading.

5. Mark this book and/or make notes in a way that makes sense to you. Always study with a pencil in your hand. Look for any kind of pattern or relationship that will help you mark, take notes, and remember what you have studied. Then, review by rereading the explanations in Chapter Three or in your notes.

6. Keep a list of the questions you have studied and when you studied each one. This plan keeps you from studying any one set of questions more than another, and it also shows you which questions you haven't studied.

7. Review often. Review after each session—immediate play-back is the best review to make information stay in your memory. Quickly review your notes before each new study session—this will establish a pattern as well as help you remember details. Conduct a major review each weekend—(1) look over your check list at the entire list of questions, (2) look at the questions you are having problems with, and (3) inspect the pattern of your study.

8. Learn more of the basics by reading automotive books and magazines. Being a Parts Specialist requires a broad knowledge—everything you read about the various tasks will help you on the test.

Pass the test by following some proven techniques:

1. When the test is distributed, do not start immediately. Scan the entire test. This gives you an idea of which items have been included.

2. Train yourself to read each question carefully. Because you are so familiar with the tasks, you can easily misread a question. Read rapidly but critically—could the question mean anything other than your first impression? The test is not intentionally tricky, but it is specific.

3. Budget your time. There are 65 questions on the test, and you have approximately two hours to complete the exam. If you find a question you don't immediately know the answer to, skip it and come back to it later. Be sure to leave a blank on the answer sheet.

4. Leave some time at the end to look over your answers. Check your answers against the answer sheet—make sure you haven't accidentally skipped a question. Only change your first answer for a good and definite reason—and only if you know your first answer is wrong. If in doubt, let your first answer stand.

CHAPTER 2 – SAMPLE QUESTIONS

1. Parts Specialist A says a PCV valve stuck in the open position can cause a rough idle.

    Parts Specialist B says a PCV valve stuck in the closed position can cause the crankshaft seals to leak. Who is correct?

    (A) A only
    (B) B only
    (C) Both A and B
    (D) Neither A nor B

2. Parts Specialist A says the air pump directs air into the exhaust manifold.

    Parts Specialist B says the air pump directs air to the catalytic converter. Who is correct?

    (A) A only
    (B) B only
    (C) Both A and B
    (D) Neither A nor B

3. Parts Specialist A says burned check valves in the air pump system can cause excessive exhaust noise.

    Parts Specialist B says burned check valves in the air pump system can cause damage to the hoses. Who is correct?

    (A) A only
    (B) B only
    (C) Both A and B
    (D) Neither A nor B

4. A gearbox that contains both a transmission and differential is a:
   - (A) transverse.
   - (B) transaxle.
   - (C) cluster gear.
   - (D) transfer case.

5. Display shelves should be:
   - (A) full to indicate the store is able to meet customer demand.
   - (B) half full to indicate high demand for the product.
   - (C) non-adjustable and of a fixed height.
   - (D) oriented parallel to the parts counter.

6. An impulse item is:
   - (A) something the customer did not originally intend to buy.
   - (B) usually stocked behind the parts counter.
   - (C) not a significant amount of the total sales.
   - (D) All of the above

7. Which of the following would NOT be considered a seasonal item?
   - (A) Battery
   - (B) Snow blower parts
   - (C) Brake shoes
   - (D) Refrigerant

8. Parts Specialist A says a 5L engine has a displacement of 350 cubic inches.

   Parts Specialist B says a thermostat marked 192 F is using metric units. Who is correct?
   - (A) A only
   - (B) B only
   - (C) Both A and B
   - (D) Neither A nor B

9. Which of these numbers would appear first in an alphanumeric listing?
   (A) 369482A
   (B) 378654C
   (C) 369482B
   (D) 388426A

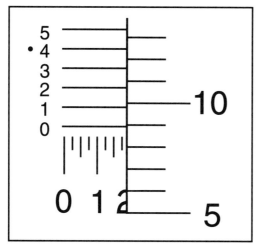

Figure 2-1 Micrometer reading.

10. The 0-1 inch micrometer reading shown in Figure 2-1 is:
    (A) 0.108
    (B) 0.184
    (C) 0.255
    (D) 0.288

11. Parts Specialist A says the best way to discourage shoplifting is to reduce the opportunity.

    Parts Specialist B says the shoplifter usually wants attention, while the customer does not. Who is correct?
    (A) A only
    (B) B only
    (C) Both A and B
    (D) Neither A nor B

12. Parts Specialist A says the do-it-yourself customer often needs help in selecting the correct parts.

    Parts Specialist B says a store with limited display area will cater to the professional technician. Who is correct?

    (A) A only
    (B) B only
    (C) Both A and B
    (D) Neither A nor B

13. Parts Specialist A says a professional technician may need help identifying the vehicle being worked on.

    Parts Specialist B says most do-it-yourself customers have had no formal automotive instruction. Who is correct?

    (A) A only
    (B) B only
    (C) Both A and B
    (D) Neither A nor B

14. Parts Specialist A says a 10 percent discount for $25.00 is $2.50.

    Parts Specialist B says a part bought for $10.00 and sold for $15.00 will generate a 50 percent profit. Who is correct?

    (A) A only
    (B) B only
    (C) Both A and B
    (D) Neither A nor B

15. A Parts Specialist:

    (A) should argue with the customer to be sure he/she buys the right parts.
    (B) has no need to understand the function of the parts.
    (C) should not recommend other available parts or services.
    (D) should be able to discuss the differences among parts lines.

16. Parts Specialist A says the best way to make sure he/she is selling the correct part is to get as much information from the customer as possible.

    Parts Specialist B says there is no need to explain the store's refund policy until the customer brings the part back. Who is correct?

    (A) A only
    (B) B only
    (C) Both A and B
    (D) Neither A nor B

17. Parts Specialist A says it is not necessary to match the customer's signature on the receipt to the signature block on the back of a credit card.

    Parts Specialist B says it is advisable to count a large amount of change very carefully. Who is correct?

    (A) A only
    (B) B only
    (C) Both A and B
    (D) Neither A nor B

18. The Parts Specialist should fill out a/an _____ for each customer purchase.

    (A) purchase order
    (B) stock order
    (C) invoice
    (D) All of the above

19. What can be done to improve the image of a counter person?

    (A) Keep a shop towel handy to clean the counter of grease and dirt from used parts.
    (B) Keep the counter free of notes, receipts, and small parts.
    (C) Keep the displays well organized and stocked.
    (D) All of the above

20. Parts Specialist A says a good understanding of vehicle systems will aid in selling the customer the correct parts.

    Parts Specialist B says suggesting related parts and components should be avoided. Who is correct?

    (A) A only
    (B) B only
    (C) Both A and B
    (D) Neither A nor B

21. Which is NOT a method used to deter shoplifters?

    (A) Congregate store personnel in one area so they can watch all directions at once.
    (B) Stagger lunch hours and breaks.
    (C) Keep displays and shelves low to improve visibility.
    (D) Keep racks and shelves full and well organized so any missing items can be easily spotted.

22. Which term describes merchandise ordered from the supplier but not shipped?

    (A) Back order
    (B) Emergency order
    (C) Stock order
    (D) Purchase order

23. Which of the following is NOT an item related to the sale of brake parts?

    (A) Brake tools
    (B) Vacuum break
    (C) Wheel bearings
    (D) Jack stands

24. Which of the following is NOT the responsibility of the Parts Specialist?

    (A) To diagnose the problem with the customer's car

    (B) To have product knowledge in order to recommend the part which best fits the customer's needs

    (C) To promote the services provided by the store and suppliers

    (D) To keep the store well stocked and orderly

25. How should the Parts Specialist react when the customer objects to buying a suggested product?

    (A) Immediately go to another customer.

    (B) Explain that the part has the best price available.

    (C) Tell the customer that the part will soon be discontinued.

    (D) Make sure the customer understands the benefits of the product.

26. If a customer returns a $100.00 part and is charged a 5 percent restocking fee, how much money is returned to the customer?

    (A) $ 5.00

    (B) $ 95.00

    (C) $ 100.00

    (D) $ 105.00

27. A customer needs 5 liters of oil for an oil change. How many quarts should the customer buy?

    (A) 2

    (B) 3

    (C) 4

    (D) 5

28. Which items can be promoted by the Parts Specialist?

    (A) Store services

    (B) Upgraded products

    (C) Sale of technical information

    (D) All of the above

29. Which is NOT a responsibility of the Parts Specialist?

    (A) Selling the customer the correct part

    (B) Solving the customer's problem with vehicle parts

    (C) Making sure every customer buys something

    (D) Creating a pleasant shopping environment for the customer

30. Closing a sale refers to:

    (A) having the customer refuse to buy.

    (B) getting the customer to commit to buy something.

    (C) taking down the sale signs.

    (D) giving up on selling to a customer.

31. Parts Specialist A says an offer to deliver the product can be used to close the sale.

    Parts Specialist B says a good salesperson will make more than one offer before accepting a lost sale. Who is correct?

    (A) A only

    (B) B only

    (C) Both A and B

    (D) Neither A nor B

32. The water pump is located:

    (A) on the engine block.

    (B) in the radiator.

    (C) in the thermostat housing.

    (D) in the engine oil pan.

33. Parts Specialist A says thermostats should be fully open at their rated temperature.

    Parts Specialist B says a thermostat opening at too low a temperature can cause engine overheating. Who is correct?

    (A) A only
    (B) B only
    (C) Both A and B
    (D) Neither A nor B

34. Which part is used to clean dirt from the air before it enters the intake manifold?

    (A) Intercooler
    (B) Oil filter
    (C) Air filter
    (D) Windscreen

35. Greeting the customer will:

    (A) help increase sales.
    (B) help prevent shoplifting.
    (C) assure the customer that helping him/her is a priority.
    (D) All of the above

36. Parts Specialist A says there is no need to get the customer's name when selling parts by phone.

    Parts Specialist B says it is often necessary to balance the time between telephone and walk-in customers. Who is correct?

    (A) A only
    (B) B only
    (C) Both A and B
    (D) Neither A nor B

37. Parts Specialist A says that the best way to sell the correct part is to get adequate information from the customer.

   Parts Specialist B says he/she should never make an assumption about a vehicle when locating parts. Who is correct?

   (A) A only
   (B) B only
   (C) Both A and B
   (D) Neither A nor B

38. Parts Specialist A says that parts catalogs usually have similar product lines grouped together.

   Parts Specialist B says the table of contents can be used to determine which catalog is needed. Who is correct?

   (A) A only
   (B) B only
   (C) Both A and B
   (D) Neither A nor B

Figure 2-2 Air conditioning system components.

39. An illustrated parts catalog will show pictures of:

    (A) where the parts are located on the vehicle.

    (B) where the parts are located in the parts bins.

    (C) the individual parts.

    (D) All of the above

40. What component in the air conditioning system is the Parts Specialist pointing to in Figure 2-2?

    (A) Accumulator

    (B) Compressor

    (C) Condenser

    (D) Evaporator

41. Parts Specialist A says that in order to buy refrigerant, the customer must show proof of successfully completing training in refrigerant recycling and service procedures.

    Parts Specialist B says the same equipment is used to service R-12 and R-134a systems. Who is correct?

    (A) A only

    (B) B only

    (C) Both A and B

    (D) Neither A nor B

42. What are three types of protection against excessive current flow?

    (A) Fuses, circuit breakers, and rectifiers

    (B) Fusible links, circuit breakers, and capacitors

    (C) Fuses, circuit breakers, and fusible links

    (D) Relays, rectifiers, and fuses

43.

OBSOLETE PART NUMBER LIST

Part Number	Status	New Number
2288SC	S	2288SG
2342	O	
2852	O	
2885	O	
3428SC	S	3428SG
3464SC	S	3464SG
3646	S	5839
3842SC	R	5820SG
3862	S	5858
3886	O	
3896	O	
3909SC	S	3988SG
5888	S	5889
39845E	O	

Status: O = Obsolete
R = Renumber
S = Superseded by

From the obsolete part number list above, which part number replaces 3909SC?

(A) 3428SG
(B) 3464SG
(C) 5820SG
(D) 3988SG

44. Parts Specialist A says a distributorless ignition system does not have a rotor.

    Parts Specialist B says a distributorless ignition system may not have spark plug wires. Who is correct?

    (A) A only
    (B) B only
    (C) Both A and B
    (D) Neither A nor B

45. Which of these is NOT a part of the exhaust system?

    (A) Muffler
    (B) Resonator
    (C) Catalytic converter
    (D) Rectifier

46. The pipe that connects the two exhaust pipes on a dual exhaust system is a:

    (A) crossover.
    (B) header.
    (C) resonator.
    (D) manifold.

47. Parts Specialist A says an example of a lost sale is a customer requesting a part that the store does not stock.

    Parts Specialist B says lost sales can include parts the store stocks that are not the brand requested by the customer. Who is correct?

    (A) A only
    (B) B only
    (C) Both A and B
    (D) Neither A nor B

48. Parts Specialist A says most modern parts stores use a computer to track inventory flow.

    Parts Specialist B says prior to computers, a card system was used to track inventory. Who is correct?

    (A) A only
    (B) B only
    (C) Both A and B
    (D) Neither A nor B

49. What part connects a transaxle to the drive wheels?

    (A) Countershaft
    (B) Line axle
    (C) Line shaft
    (D) Halfshaft

50. What part connects a driveshaft to the pinion yoke of the differential?

    (A) Universal joint (U-joint)
    (B) Pinion joint
    (C) Halfshaft
    (D) Pinion nut

51. A limited-slip differential:

    (A) limits the speed difference between the drive wheels.
    (B) requires a special lubricant.
    (C) uses clutch plates to limit the slip.
    (D) All of the above

52. Parts Specialist A says the brake pedal should never pulse on a vehicle equipped with anti-lock brakes.

    Parts Specialist B says anti-lock brakes shorten the braking distance and improve handling on a slippery road surface. Who is correct?

    (A) A only
    (B) B only
    (C) Both A and B
    (D) Neither A nor B

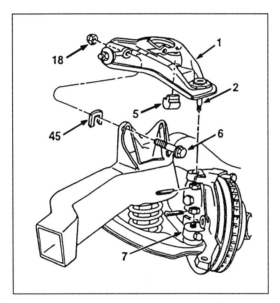

Figure 2-3   Upper control arm attachment
(Courtesy of General Motors Corporation).

53. While referring to Figure 2-3 above, Parts Specialist A says item #2 is a ball joint.

    Parts Specialist B says item #2 allows the steering knuckle to turn right and left. Who is correct?

    (A) A only
    (B) B only
    (C) Both A and B
    (D) Neither A nor B

54. On most current production automobiles, the heater air temperature inside the vehicle is controlled by:

   (A) varying the amount of engine coolant sent to the heater core.

   (B) blending incoming heated air with fresh air.

   (C) varying the temperature of the engine coolant.

   (D) cycling the air conditioner compressor to cool off the air.

55. A customer asks about the difference between standard and heavy duty turn signal flashers. Which statement is true?

   (A) A heavy duty flasher will flash faster than a standard duty flasher.

   (B) A heavy duty flasher should be used when the standard duty flasher flashes too fast.

   (C) A standard duty flasher is best when pulling trailers.

   (D) There is no difference.

56. Bolt diameter is measured by:

   (A) the distance across the flats of the hex head.

   (B) the size wrench that fits on the head.

   (C) the distance across the points of the hex head.

   (D) the diameter of the threads.

57. Parts Specialist A says the thread pitch of an American bolt is measured by the distance between the threads in inches.

   Parts Specialist B says the thread pitch of a metric bolt is measured by the distance between the threads in millimeters. Who is correct?

   (A) A only

   (B) B only

   (C) Both A and B

   (D) Neither A nor B

58. Which type of fastener does NOT have a head?

    (A) Screw

    (B) Nut

    (C) Bolt

    (D) Stud

59. Which statement does NOT describe a safe practice for lifting heavy objects?

    (A) Lift with your legs.

    (B) Keep the weight close to your body.

    (C) Make sure the walkway is clear before lifting the object.

    (D) Bend your back over the object to get a better grip.

60. Parts Specialist A says most chemicals found in a parts store can be some form of health hazard.

    Parts Specialist B says information regarding the handling of each of these chemicals can be found on its Material Safety Data Sheet (MSDS). Who is correct?

    (A) A only

    (B) B only

    (C) Both A and B

    (D) Neither A nor B

61. Which statement is true regarding hazardous waste disposal?

    (A) Avoid hazardous waste by switching to a biodegradable solvent or oil.

    (B) Neutralize caustic cleaning solutions prior to disposal.

    (C) Contract with a hazardous waste disposal company to handle hazardous waste.

    (D) All of the above

62. A bolt with three radial lines embossed on the head would be:
    - (A) grade 1.
    - (B) grade 3.
    - (C) grade 5.
    - (D) metric.

63. Three types of fittings used on automobiles are:
    - (A) flare, compression, and tubing.
    - (B) pipe, flare, and steel.
    - (C) tubing, steel, and pipe.
    - (D) flare, compression, and pipe.

64. Which type of fitting relies on tapered threads to prevent the leak?
    - (A) Pipe
    - (B) Compression
    - (C) Flare
    - (D) Tubing

65. Parts Specialist A says information regarding the proper application of a chemical can be found on the product labels.

    Parts Specialist B says information regarding the proper application of a chemical can be found on its Material Safety Data Sheet (MSDS). Who is right?
    - (A) A only
    - (B) B only
    - (C) Both A and B
    - (D) Neither A nor B

66. What is the standard location of the Vehicle Identification Number (VIN)?

    (A) On a plate attached to the floorpan under the driver seat

    (B) On the bottom of the hood

    (C) On the package tray behind the rear seat

    (D) Attached to the driver-side top of the instrument panel and visible through the windshield

67. Which information is NOT available from the Vehicle Identification Number (VIN)?

    (A) Paint color

    (B) Manufacturer

    (C) Year model

    (D) Engine size

68. Parts Specialist A says the outboard Constant Velocity (CV) joint on a front-wheel drive vehicle must allow for changes in the thrust angle.

    Parts Specialist B says the inboard CV joint allows for the in-and-out motion. Who is correct?

    (A) A only

    (B) B only

    (C) Both A and B

    (D) Neither A nor B

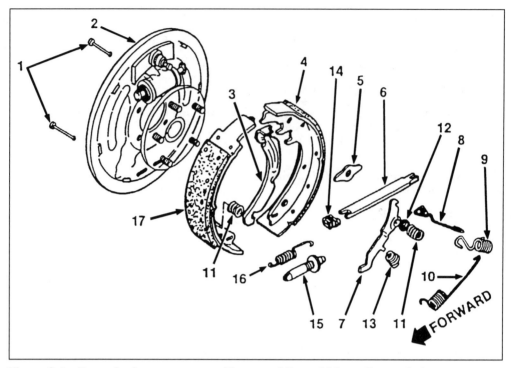

Figure 2-4 Drum brake components *(Courtesy of General Motors Corporation)*.

69. Refer to Figure 2-4 above. The part shown as item #6 is used to:

    (A) release the brakes.
    (B) apply the parking brake.
    (C) adjust the hydraulic brakes.
    (D) adjust the parking brake.

70. A parts customer buys brake pads for his vehicle. Which of the following items should NOT be suggested by the Parts Specialist?

    (A) Brake spring pliers
    (B) Brake fluid
    (C) Caliper rebuild kit
    (D) Wheel bearing grease

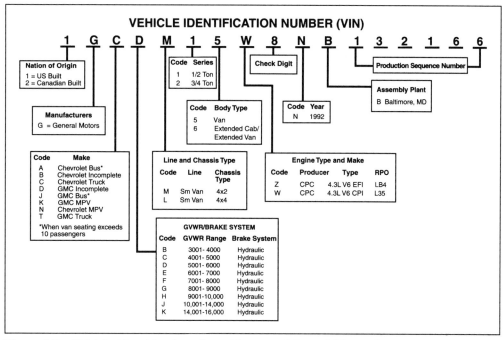

Figure 2-5 Vehicle identification chart *(Courtesy of General Motors Corporation)*.

71. Referring to the Vehicle Identification Number (VIN) in Figure 2-5, what is the GVWR rating of the vehicle?

    (A) 3001-4000

    (B) 4001-5000

    (C) 5001-6000

    (D) 6001-7000

72. Which of the following is NOT true of a core charge?

    (A) It applies to remanufactured or rebuilt parts.

    (B) It is refunded when the customer brings in a rebuildable part to replace the one purchased.

    (C) It helps reduce the purchase price of the parts.

    (D) All parts are subject to a core charge.

73. Parts Specialist A says engine pistons are available in different sizes to allow for cylinder block boring.

   Parts Specialist B says rod and main insert bearings are available in different sizes to allow for camshaft journal machining. Who is correct?

   (A) A only
   (B) B only
   (C) Both A and B
   (D) Neither A nor B

74. A thermostat is used in the engine to:

   (A) set the maximum operating temperature.
   (B) control the air flow through the radiator.
   (C) increase the engine warm-up time.
   (D) control the water flow through the radiator.

75. The engine fan:

   (A) is used to increase air flow through the radiator during low vehicle speeds.
   (B) may be mounted on the end of the water pump.
   (C) may be driven by an electric motor.
   (D) All of the above

76. Parts Specialist A says warranty-returned parts should be placed back in the inventory.

   Parts Specialist B says additional paperwork is usually required with warranty returns in order to receive credit from the manufacturer. Who is correct?

   (A) A only
   (B) B only
   (C) Both A and B
   (D) Neither A nor B

77. Which term describes a single item?
    (A) Each
    (B) Carton
    (C) Case
    (D) Pair

Figure 2-6 Transmission I.D. location.

78. In the transmission identification number in Figure 2-6, the first two numbers indicate the year of manufacture, the third and fourth numbers identify the model, and the last five numbers give the serial number. Referring to the transmission ID number (85HF 03234) in Figure 2-6, when was this transmission produced?
    (A) 1992
    (B) 1982
    (C) 1985
    (D) 1975

79. The part which transfers the fuel from the tank to the carburetor is the:

    (A) fuel pump.

    (B) fuel injector.

    (C) fuel regulator.

    (D) fuel rectifier.

80. The oxygen sensor is located:

    (A) between the exhaust manifold and the catalytic converter.

    (B) in the intake manifold.

    (C) in the air cleaner housing.

    (D) in the fuel line.

81. Parts Specialist A says component ID data is stamped on major components such as engine blocks, transmission cases, and differential housings.

    Parts Specialist B says component ID information has no value in identifying the part. Who is correct?

    (A) A only

    (B) B only

    (C) Both A and B

    (D) Neither A nor B

82.

RPO CODE LISTING	
GQ1	Rear Axle – Base Ratio
GT4	Rear Axle – 3.73 Ratio
GU5	Rear Axle – 3.23 Ratio
GU6	Rear Axle – 3.42 Ratio
G50	Heavy Duty Rear Springs. Required with X8F or X8H Payload Ratings.
G80	Rear Axle – Positraction
JA4	Power Brakes – Front Disc, Rear Drum, Right Rear Wheel Anti-lock
JM4	Power Brakes – Front Disc, Rear Drum, Front Wheels and Right Rear Wheel Anti-lock
KC4	Cooler System – Engine Oil
KO5	Heater – Engine Block
K34	Cruise Control – Automatic, Electronic
K60	Generator – 100 Amp
K-99	Generator – 85 Amp
LB4	Engine – Gas 6 Cylinder, 4.3L TBI
L35	Engine – Gas 6 Cylinder, 4.3L CPI

Parts Specialist A says an RPO code of GU6 means the vehicle has a 3.73 rear axle ratio.

Parts Specialist B says an RPO code of K60 means the vehicle is equipped with an 85 amp generator. Who is correct?

- (A) A only
- (B) B only
- (C) Both A and B
- (D) Neither A nor B

83. Parts Specialist A says a hatchback body style has a rear glass that opens to allow access to the luggage compartment.

    Parts Specialist B says a fastback body style has a sloping rear glass that does not lift for access to the luggage compartment. Who is correct?

    - (A) A only
    - (B) B only
    - (C) Both A and B
    - (D) Neither A nor B

84. Which emissions system removes hydrocarbons from the engine's crankcase?

    - (A) Air Injection Reaction (AIR)
    - (B) Exhaust Gas Recirculation (EGR)
    - (C) Early Fuel Evaporization (EFE)
    - (D) Positive Crankcase Ventilation (PCV)

85. Which is NOT a pollutant produced by the engine?

    - (A) Carbon dioxide
    - (B) Carbon monoxide
    - (C) Hydrocarbons
    - (D) Oxides of nitrogen

86. Which emissions system is used to reduce oxides of nitrogen?

    (A) Air Injection Reaction (AIR)

    (B) Exhaust Gas Recirculation (EGR)

    (C) Early Fuel Evaporization (EFE)

    (D) Positive Crankcase Ventilation (PCV)

87. Which of the following are sources of information in addition to the parts catalog?

    (A) Service manuals

    (B) Service bulletins

    (C) Supplier training materials

    (D) All of the above

88. Parts Specialist A says the paint code is located on the Vehicle Identification Number (VIN) tag.

    Parts Specialist B says the paint code is on the trim tag. Who is correct?

    (A) A only

    (B) B only

    (C) Both A and B

    (D) Neither A nor B

89. Parts Specialist A says the pilot bearing is on the end of the clutch-release lever.

    Parts Specialist B says the pilot bearing is located in the center of the flywheel. Who is correct?

    (A) A only
    (B) B only
    (C) Both A and B
    (D) Neither A nor B

90. What devices are used to engage the various gear ranges of an automatic transmission?

    (A) Bands
    (B) Clutches
    (C) Servos
    (D) All of the above

91. Parts Specialist A says a torque converter in an automatic transmission replaces the clutch plate found in a manual transmission.

    Parts Specialist B says a torque converter provides additional gear reduction during periods of acceleration. Who is correct?

    (A) A only
    (B) B only
    (C) Both A and B
    (D) Neither A nor B

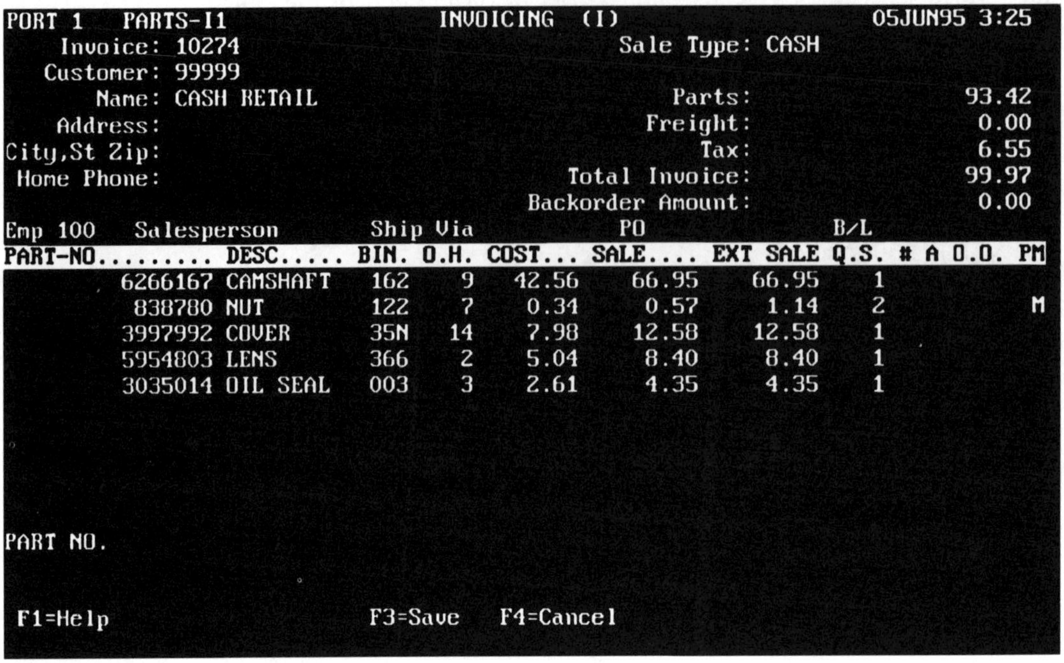

```
PORT 1    PARTS-I1                  INVOICING  (I)                    05JUN95 3:25
     Invoice: 10274                              Sale Type: CASH
    Customer: 99999
         Name: CASH RETAIL                          Parts:                93.42
      Address:                                    Freight:                 0.00
 City,St Zip:                                         Tax:                 6.55
   Home Phone:                                Total Invoice:              99.97
                                           Backorder Amount:               0.00
Emp 100    Salesperson         Ship Via           PO              B/L
PART-NO......... DESC..... BIN. O.H. COST... SALE.... EXT SALE Q.S. # A O.O. PM
        6266167 CAMSHAFT    162   9   42.56    66.95    66.95   1
        838780  NUT         122   7    0.34     0.57     1.14   2              M
        3997992 COVER       35N  14    7.98    12.58    12.58   1
        5954803 LENS        366   2    5.04     8.40     8.40   1
        3035014 OIL SEAL    003   3    2.61     4.35     4.35   1


PART NO.

F1=Help                 F3=Save    F4=Cancel
```

Figure 2-7 Customer invoice.

92. According to Figure 2-7, what is the total amount the customer should pay?

   (A) $ 6.55
   (B) $ 66.95
   (C) $ 93.42
   (D) $ 99.97

93.

PARTS INTERCHANGE CHART		
Repl. No.	Mfg.	Part No.
AB1194	ABCD	52400
AB1208	ABCD	632
AB1209	ABCD	633
AB1210	ABCD	742
AB1211	ABCD	1328
AB1213	ABCD	02820
AB1214	ABCD	3320
AB1215	ABCD	3420
AB1220	ABCD	15244
AB1221	ABCD	15245
AB1223	ABCD	1872
AB1224	ABCD	24721

What part number will replace an AB1220 manufactured by the ABCD company?

(A) 15244
(B) 1220
(C) 14276
(D) 15245

94. A customer says a car's battery voltage is higher than specified with the engine running. What part is causing the problem?

(A) Rectifier
(B) Diodes
(C) Ignition coil
(D) Voltage regulator

95. Parts Specialist A says a defective alternator can cause the battery not to be charged.

   Parts Specialist B says a defective voltage regulator can cause the battery not to be charged. Who is correct?

   (A) A only
   (B) B only
   (C) Both A and B
   (D) Neither A nor B

96. Parts Specialist A says cars built in the United States use only American National Standard (English) threads.

   Parts Specialist B says metric threads are used only on Japanese manufactured vehicles. Who is correct?

   (A) A only
   (B) B only
   (C) Both A and B
   (D) Neither A nor B

97.
   COIL SPRING IDENTIFICATION CHART

I.D.	Wire Diameter	Number of Coils	Load	Part Number
3.430	.468	10.00	500	399-5363
3.881	.500	9.50	443	399-5095
5.036	.918	11.00	3330	399-1563
5.036	.563	8.00	1306	399-1561
5.395	.593	10.00	380	399-5333

   From the coil spring identification chart above, which part number spring has an I.D. of 5.036 and a wire diameter of 0.563?

   (A) 399-5095
   (B) 399-1563
   (C) 399-1561
   (D) 399-5333

98.

ABBREVIATION CHART		
Alum	....	Aluminum
BF	....	Barrel Face
cc	....	Cubic Centimeter
Chr	....	Chrome
CI	....	Cast Iron
Cm	....	Compression Ring
Cyl	....	Cylinder
Dia	....	Diameter
Dist	....	Distilled
DOHC	....	Dual Overhead Cam
Duct	....	Ductile Iron
Eng	....	Engine
Exc	....	Except
Gr	....	Grooved
H/O	....	High Output
H.P.	....	Horsepower

From the abbreviation chart above, what does the abbreviation *Chr* represent?

(A) Chrysler
(B) Chromate
(C) Chrome
(D) Chromium

99.

TABLE OF CONTENTS

	Page No.
Introduction	
• Table of Contents	A
• How to Use this Publication	B
• Key to Abbreviations	1
Section One (Bearing Interchanges)	3-349
Section Two (Bearing Specifications)	
General Information	
• Specification Guide	350
• Bearing Types	351
• Bearing Sets with Competitor Interchange	352
I. Tapered Roller Bearings	
• Descriptions and Illustrations	353-354
• Numerical Listing	355-388
• Size Listing by Bore Size	389-433
II. Cylindrical Roller Bearings	
• Descriptions and Illustrations	434-440
• Numerical Listing	441-443
• Size Listing by Bore Size	444-450

Referring to the table of contents above, on which page would the Parts Specialist expect to find an illustration of a tapered roller bearing?

- Ⓐ Page 300
- Ⓑ Page 352
- Ⓒ Page 353
- Ⓓ Page 435

100. Parts Specialist A says supersession bulletins refer to catalog errors due to typing errors or inaccurately assigned part numbers.

   Parts Specialist B says technical bulletins note part numbers that now supersede previous part numbers. Who is correct?

   (A) A only
   (B) B only
   (C) Both A and B
   (D) Neither A nor B

101. Parts Specialist A says to adjust the cable if the parking brake will not hold.

   Parts Specialist B says checking the correct brake fluid level in the master cylinder is required maintenance for proper parking brake operation. Who is correct?

   (A) A only
   (B) B only
   (C) Both A and B
   (D) Neither A nor B

102. The control arm on an SLA (Short-Long Arm) suspension system connects to the frame through the:

   (A) ball joint.
   (B) coil spring.
   (C) idler arm.
   (D) bushings.

103. Which part is used with a rack-and-pinion steering system?

   (A) Tie-rod end
   (B) Drag link
   (C) Idler arm
   (D) Pitman arm

104. What are the two types of steering gearboxes?

  (A) Recirculating ball worm and MacPhearson
  (B) MacPhearson and rack-and-pinion
  (C) Recirculating ball worm and rack-and-pinion
  (D) Pitman and MacPhearson

105.

BIN	PART NUMBER	1ST-CNT	2ND-CNT	NEW BIN
35N	481550 MLDG KITC	( )	( )	( )
35N	2008004 GEAR	( )	( )	( )
35N	3698741 SUPPORT	( )	( )	( )
35N	3774720 CLUTCH	( )	( )	( )
35N	3793659 LEVER	( )	( )	( )
35N	3867815 GEAR	( )	( )	( )
35N	3885124 GSKT	( )	( )	( )
35N	3893719 RUN	( )	( )	( )
35N	3915034 GEAR ASM	( )	( )	( )
35N	3930800 GASKET	( )	( )	( )
35N	3962943 BEZEL LH	( )	( )	( )
35N	3962944 BEZEL RH	( )	( )	( )
35N	3973863 CABLE ASM	( )	( )	( )
35N	3974217 ARM & BALL	( )	( )	( )

What would be the purpose of this computer listing?

  (A) Physical inventory
  (B) Invoicing
  (C) Stock order
  (D) Cataloging

106.

LOCATOR-PAD		PAGE NO: 1	
ADP MOTORS		DATE: 05 JUN 1995	
S.O.	PART-NO.	ON-HAND	BIN
105	325203	2	UP
107	325730	1	UP
102	331248	3	U43
101	332395	1	785
102	332396	0	741
107	334360	11	644
101	336013	1	102
101	33924A	3	107
100	339654	2	SHED
102	339669	0	107
100	340190	1	512
101	340194	0	334
101	345806	2	387
105	355798	5	UP
107	394678	4	245
102	399571	4	429
100	451875	4	150
104	457196	1	455
101	481540	1	PL
102	481550	1	35N
101	485019	1	BLK
100	642871	2	427

Referring to the above listing, Parts Specialist A says there are two parts numbered 355798 in stock.

Parts Specialist B says part number 332395 is located in bin 741. Who is correct?

- (A) A only
- (B) B only
- (C) Both A and B
- (D) Neither A nor B

107. Which of the following should NOT be done if an inventory discrepancy is noticed?

   (A) Ignore the problem.
   (B) Verify the discrepancy.
   (C) Make a note of the part numbers affected.
   (D) Notify your supervisor.

108. *Stock rotation* refers to:

   (A) rotating the product so the labels are facing the front of the shelf.
   (B) selling the older stock first.
   (C) selling the newer stock first.
   (D) rearranging the shelves for a neater appearance.

109. Parts Specialist A says a special order is a non-stocking part which is ordered for a customer.

   Parts Specialist B says records of special orders and lost sales can indicate which parts need to be added to the inventory. Who is correct?

   (A) A only
   (B) B only
   (C) Both A and B
   (D) Neither A nor B

110. Parts Specialist A says a port fuel injection system has an injector for each cylinder.

   Parts Specialist B says a throttle body injection system has one or two injectors for the entire engine. Who is correct?

   (A) A only
   (B) B only
   (C) Both A and B
   (D) Neither A nor B

111. The heat range of a spark plug indicates:

    (A) whether the plug is designed for summer or winter driving.

    (B) how quickly the plug transfers heat to the head.

    (C) what size coil is required to fire the plug.

    (D) the expected life of the plug.

112. The voltage needed to create the spark for the spark plug is produced by the:

    (A) distributor.

    (B) rotor.

    (C) alternator.

    (D) ignition coil.

113. The term *core charge* refers to:

    (A) the basic price of an item without any accessories.

    (B) an additional charge added to a remanufactured part.

    (C) the cost of shipping the part to the store.

    (D) the cost of handling a returned part.

114. Parts Specialist A says a set refers to parts packaged in the quantity needed to complete the repair.

    Parts Specialist B says to purchase a quantity less than a set usually requires a special order. Who is correct?

    (A) A only

    (B) B only

    (C) Both A and B

    (D) Neither A nor B

115. How can the Parts Specialist handle a broken kit?
    (A) Sell it as a complete unit.
    (B) Dispose of it since it has no resale value.
    (C) Order replacement parts so it can be sold once it is complete.
    (D) All of the above

116. The crankshaft is held in the engine block by the:
    (A) main bearing caps.
    (B) rod bearing caps.
    (C) cylinder head.
    (D) shaft retainer clip.

117. Camshafts may be driven by:
    (A) timing gears.
    (B) timing chains.
    (C) timing belts.
    (D) All of the above

118. A customer is purchasing a head gasket for an overhead cam engine. Which additional part(s) should the Parts Specialist suggest?
    (A) Timing belt
    (B) Hydraulic lifters
    (C) Pushrods
    (D) All of the above

119. Parts Specialist A says there is no need to make a note of customer exchanges if the price is the same on the switched parts.

   Parts Specialist B says the only adjustment needed on a returned part is to refund the customer's money. Who is correct?

   (A) A only
   (B) B only
   (C) Both A and B
   (D) Neither A nor B

120. In which location is merchandise most likely to get the attention of the customer?

   (A) On the top shelves
   (B) At eye level
   (C) On the bottom shelves
   (D) Stacked on the floor

121. High-volume sales items should be located:

   (A) close to the cash register.
   (B) close to the door.
   (C) in the center of the store.
   (D) along the walls and away from the entrance.

122. Individual pricing should be used for:

   (A) parts in the stock room.
   (B) items displayed in front of the parts counter.
   (C) parts sold in less than case quantities.
   (D) all parts regardless of their location.

123. Parts Specialist A says it is better to ask for help than to sell the customer the wrong part.

Parts Specialist B says an experienced employee may be asked to help train a new employee. Who is correct?

(A) A only

(B) B only

(C) Both A and B

(D) Neither A nor B

124. A Parts Specialist is responsible for keeping the _____ clean and orderly.

(A) floors

(B) shelves

(C) displays

(D) All of the above

125. Parts Specialist A says battery acid can cause severe burns if it comes in contact with skin or clothing.

Parts Specialist B says there is no hazard associated with charging a battery. Who is correct?

(A) A only

(B) B only

(C) Both A and B

(D) Neither A nor B

126. What should the Parts Specialist do when a phone inquiry interrupts serving a walk-in customer?

(A) Answer the phone and ask the caller to hold.

(B) Ignore the phone and let the caller call back later.

(C) Tell the walk-in customer to wait while the Specialist handles the phone call.

(D) Handle the walk-in and phone customer at the same time to keep everybody happy.

127. Which person is most likely to sell machine shop services to the customer?

- (A) Store owner
- (B) Machinist
- (C) Shop owner
- (D) Counterperson

128. Which sandpaper has the finest grit?

- (A) 40
- (B) 80
- (C) 400
- (D) 800

129. Parts Specialist A says fuel hose and vacuum hose are interchangeable.

    Parts Specialist B says vacuum hose can be used on the windshield washer system. Who is correct?

- (A) A only
- (B) B only
- (C) Both A and B
- (D) Neither A nor B

130. Parts Specialist A says copper and steel tubing is measured by the outside diameter.

    Parts Specialist B says vacuum hose is measured by the inside diameter. Who is correct?

- (A) A only
- (B) B only
- (C) Both A and B
- (D) Neither A nor B

CHAPTER 3 – ANSWERS AND ANALYSIS

1. The correct answer is **C**.

   Both Specialists are correct. A PCV valve stuck in the open position will create a vacuum leak and, thereby, a rough idle. If it is stuck in the closed position, the blowby from the cylinders cannot escape. This will allow pressure to build in the crankcase and cause oil leakage past the crankcase seals.

2. The correct answer is **C**.

   Both Specialists are correct. The air pump directs air to the exhaust manifold during cold engine operation. This allows the air to mix with the hot exhaust gases as they leave the engine. Adding extra air to the exhaust gases makes possible the continued burning of any hydrocarbon and carbon monoxide remaining in the exhaust. After the engine is up to operating temperature, air is injected into the catalytic converter. This extra air is used by the converter to change oxygen, hydrocarbons, and carbon monoxide into water vapor and carbon dioxide. If the air is injected into the exhaust manifold after the engine is in closed-loop (computer controlled) operation, the extra air will cause the oxygen sensor to produce inaccurate signals.

3. The correct answer is **C**.

   Burned check valves in the air pump system can allow exhaust to blow out under the hood, creating excessive noise. The rubber hoses between the air pump and check valves will be burned if the hot exhaust gas gets past the valves.

4. The correct answer is **B**.

   A **transaxle** is a combination transmission and differential. *Transverse* refers to the engine and transaxle being mounted crossways in the vehicle. A *cluster gear* is a part of a manual shift transmission. A *transfer case* connects the front and rear drivelines of a four-wheel drive vehicle.

5. The correct answer is **(A)**.

Display shelves should be **kept full to indicate the store is able to meet customer demand** (see Figure 3-1). Half-full or empty shelves create customer doubts about the product. The shelving should be adjustable to accommodate various sizes of merchandise. The display shelves should run perpendicular to the parts counter to allow better eye contact with the customers when answering their questions and also to discourage shoplifters.

6. The correct answer is **(A)**.

An impulse item is **something the customer did not intend to buy** when entering the store. Impulse items are usually stocked in front of the parts counter (see Figure 3-2) and can be a significant amount of the total sales.

7. The correct answer is **(C)**.

Brake shoes sell equally well year round. Batteries and snow blower parts are seasonal items as they have higher sales in the winter. Refrigerant is also seasonal as it is mainly sold during the spring and summer months.

8. The correct answer is **(D)**.

Neither Specialist is correct. There are 61 cubic inches to a liter. To convert liters to cubic inches, multiply the liters by 61 to get cubic inches. Following this equation, 5L x 61 equals 305 cubic inches, not 350. The F following the number signals Fahrenheit. Most engines use thermostats calibrated in the range of 180 to 192 degrees Fahrenheit (F). The metric unit of temperature measurement is degrees Celsius (C). When making the conversion, 192 degrees Fahrenheit is equal to 89 degrees Celsius.

9. The correct answer is **(A)**.

The first number would be **369482A**. An alphanumeric listing places the numbers in order starting from the left digit and working across to the right. The two 36***** numbers would be before the 37***** and 38*****. Both of the 369482 numbers are the same with the letter A coming before C.

Figure 3-1 Display shelves should be kept full.

Figure 3-2 Impulse items are usually stocked near the checkout area.

10. The correct answer is **B**.

Because this a 0-1 inch micrometer, the reading must be between 0 and 1 inch:	0.000
The 1 is visible on the horizontal line:	0.100
Three marks are visible after the 1 (3 x 0.025):	0.075
The horizontal line nearly lines up with the 9 on the vertical line:	0.009
The total measurement is:	**0.184**

Figure 3-3 Display shelves should be easily visible from checkout areas.

11. The correct answer is **A**.

Parts Specialist A is correct. The best way to avoid shoplifting is to reduce the opportunity. Keep small, expensive items in a locked cabinet. Display large items near the exit door and small, easily stolen items at the end of the store opposite the exit door. Arrange display shelves perpendicular to checkout areas (see Figure 3-3). The shoplifter tries to **avoid** attention, whereas the paying customer wants attention and service.

12. The correct answer is **C**.

    Both Specialists are correct. The do-it-yourself customer is often unsure of the proper identification of the vehicle and needs help selecting the parts that are needed. Professional technicians are not usually interested in shopping. They prefer a parts counter with plenty of help so they can get their parts and be back in the shop in a minimum amount of time.

13. The correct answer is **B**.

    Parts Specialist B is correct—most do-it-yourself customers have not received any formal automotive repair training. In contrast, most professional technicians know the importance of positively identifying the vehicle being worked on and will bring that information to the store.

14. The correct answer is **C**.

    Both Specialists are correct. To calculate percent discount:
    (1) Convert the percent into a decimal number by moving the decimal two places to the left—10 percent becomes 0.10.
    (2) Multiply the decimal number by the original price to get the amount of the discount—0.10 x $25.00 is $2.50.

    To calculate percent profit:
    (1) Subtract the cost from the selling price—$15.00 minus $10.00 is $5.00.
    (2) Divide this difference by the cost—$5.00 divided by $10.00 is 0.5.
    (3) Convert this number into a percentage by moving the decimal two places to the right—0.5 becomes 50 percent.

15. The correct answer is **D**.

    The Parts Specialist should be able to **discuss the differences among parts and parts lines**. The customer may ask how the parts differ in order to make a better decision about which parts to buy. It is not the Specialist's place to force an opinion on the customer. By understanding the basic function of the parts, the Specialist is able to determine if the proper part is being purchased and if any related parts are needed. He/she should also make the customer aware of any other parts and services the store may offer that the customer needs.

16. The correct answer is (A).

    Parts Specialist A is correct. He/she should get as much information as necessary to be sure to sell the customer the correct part. The customer should also be aware of the return policy regarding the part before it is purchased. The customer should not be surprised with no-return policies and restocking fees when returning a part.

17. The correct answer is (B).

    Specialist B is correct. If a customer pays by credit card, always verify the signature on the receipt to the signature block on the back of the credit card. If a customer pays in cash with a large bill, count the change carefully.

18. The correct answer is (C).

    An **invoice** should be completed for each sale in the store. These invoices are used to track inventory and customer purchases for billing purposes. A *purchase order* is used to allow a company to purchase parts. It will describe the parts and quantity of parts to be purchased along with billing information. A *stock order* is used by the store to order more stock from the suppliers.

19. The correct answer is (D).

    All of these aspects—clean counters and organized, well stocked displays—will help improve the image of the parts counter and the counter personnel.

20. The correct answer is (A).

    Specialist A is correct. A good knowledge of vehicle systems will help in selling the correct part and suggesting related items and services the customer may have forgotten. Additional parts should be suggested whenever appropriate. This approach can increase sales and prevent the customer from having to come back for more parts later.

21. The correct answer is (A).

    Store personnel should not congregate in one area. Congregating personnel increases the number of unwatched areas in the store. Personnel should be dispersed throughout the store as evenly as practical in order to watch all areas. Staggering the lunch hours and breaks will maximize the number of people available to watch the store. Keeping the displays and shelves low will increase the store personnel's ability to see all areas of the store at once. Full, neat shelves will make it more obvious when parts are sold or missing. Shoplifters seek out cluttered areas where their theft is harder to detect.

22. The correct answer is **(A)**.

    Back order refers to merchandise ordered from a supplier but not shipped due to the supplier being out of stock. An *emergency order* is an order placed out of the normal cycle of stock orders. A *stock order* is an order placed with the supplier on a routine basis, usually on a weekly or biweekly basis. A *purchase order*, the company's tool to purchase parts, describes the parts and quantity of parts to be purchased along with the billing information.

23. The correct answer is **(B)**.

    A **vacuum break** is a carburetor part. Various hand tools are required to properly disassemble and reassemble the components of a brake system. Wheel bearings should be cleaned, checked, and re-packed with grease during a brake repair job. Jack stands are used to support the vehicle while the work is being performed. These tools will make the customer's brake repair job faster, easier, and safer.

24. The correct answer is **(A)**.

    Diagnosing the customer's problem or advising the customer on how to repair the vehicle is not the responsibility of the Parts Specialist. The Specialist can provide assistance by informing the customer of the benefits and features of a part or service, but the final decision rests with the customer. The Specialist should have product knowledge, promote the available services, and keep the store stocked and orderly, but be sparing with advice.

25. The correct answer is **(D)**.

    The Parts Specialist should, in a polite manner, **be sure the customer understands the benefits of the product being discussed**. Avoid such hard-sell techniques as stressing price or limited availability. Do not abandon the customer until it is obvious he/she does not wish to purchase the item.

26. The correct answer is **(B)**.

The customer receives **$95.00**, which is 95 percent of $100.00. The restocking fee, usually expressed as a percent, is the fee charged for having to handle a returned part.

To calculate the restocking fee:

(1) Convert the percent into a decimal number by moving the decimal two places to the left—5 percent becomes 0.05.

(2) Multiply the decimal number by the original price to get the amount of the restocking fee—0.05 x $100.00 is $5.00.

(3) Subtract the restocking fee from the original price to get the amount of money to be returned to the customer—$100.00 minus $5.00 is $95.00 given to the customer.

27. The correct answer is **(D)**.

The customer **should buy 5 quarts**. One quart is equal to 0.9464 liters, so one quart is almost equal to one liter. Therefore, 5 liters is approximately equal to 5 quarts. In actual practice, this amount will leave the engine 1/4 quart low on oil but within the safe operating range.

28. The correct answer is **(D)**.

All of these items—store services, upgraded products, sale of technical information—are usually promoted by the Parts Specialist.

29. The correct answer is **(C)**.

It is **not necessary that every customer visiting the store make a purchase**. Solving the customer's problem correctly and creating a pleasant shopping environment will motivate the customer to return.

30. The correct answer is **(B)**.

Closing a sale means **getting the customer to commit to buy something**. Having the customer refuse to buy or giving up on the customer both describe a lost sale.

31. The correct answer is **(C)**.

Both Specialists have made good suggestions. Many techniques can be used in order to close a sale. These offers can include delivery, deferred billing, or additional accessories. A good salesperson will use at least one of these techniques before accepting a lost sale.

32. The correct answer is **(A)**.

    The water pump is usually located **on the engine block**.

33. The correct answer is **(D)**.

    Neither Parts Specialist is correct. A thermostat will begin to open at its rated temperature and should be fully open at a 20°F higher temperature. If a thermostat opens at too low a temperature, the engine will run too cold.

34. The correct answer is **(C)**.

    The **air filter** removes dirt from the air. An intercooler is used with a turbocharger or supercharger to cool the air entering the intake manifold. The oil filter removes dirt from the oil. "Windscreen" is a British term referring to the windshield.

35. The correct answer is **(D)**.

    Greeting the customer as he/she enters the store will help in **all of these areas:** increasing sales, preventing shoplifting, and assuring the customer that helping him/her is a top priority.

36. The correct answer is **(B)**.

    Parts Specialist B is correct—good customer relations often depends on the ability to balance the time between telephone and walk-in customers. In the other situation, a Specialist should also get the customer's name when selling parts by phone. These notes can often be used for reference when the customer comes in to purchase the parts.

37. The correct answer is **(C)**.

    Both Specialists are right. Adequate and correct information is the only way to look up parts accurately for a vehicle. In addition, the Parts Specialist should never make assumptions about how a vehicle is equipped.

Figure 3-4 The table of contents of each product catalog will make a quick examination easier.

38. The correct answer is **C**.

    Both Parts Specialists are correct. Similar product lines are usually grouped together in a catalog. In addition, the table of contents always makes a quick examination possible (see Figure 3-4).

39. The correct answer is **C**.

    An illustrated parts catalog will show pictures of the **individual parts** for identification purposes. The location of the parts on the vehicle is usually shown in the service manuals. As all parts bins can be arranged differently, no manual will show where the parts are located in the stock room.

40. The correct answer is **D**.

    The part indicated is an **evaporator**. It is positioned in the system between the orifice tube and the accumulator. The evaporator looks similar to a heater core and is located in the passenger compartment. The orifice tube causes the drop in refrigerant pressure that results in the change in temperature necessary for cooling. The *accumulator* is a round, metal can mounted in the engine compartment close to the evaporator. It contains a desiccant, which absorbs moisture from the refrigerant. The *compressor*, located on the engine and driven by a belt, pumps the refrigerant through the system. The *condenser* looks similar to a radiator and is located at the front of the vehicle, usually in front of the radiator.

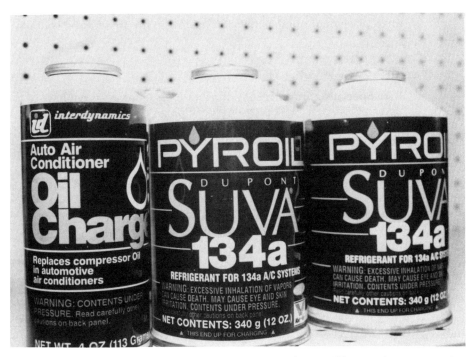

Figure 3-5 Proof of training must be shown to purchase refrigerant in any quantity.

41. The correct answer is **A**.

    In order for any customer to purchase refrigerant (see Figure 3-5), he/she **must show proof of having successfully completed training in refrigerant recycling and service procedures**. This proof is required to buy refrigerant in any quantity. Separate equipment must be purchased to service R-12 and R-134a systems.

42. The correct answer is **C**.

    The three devices used to protect a circuit from excessive current flow are **fuses, circuit breakers, and fusible links**. These devices protect the circuit by interrupting the current flow when it exceeds their rated capacity. Fuses and fusible links must be replaced, while circuit breakers may be of the manual or automatic reset type. Rectifiers convert alternating current (AC) into direct current (DC). Capacitors are used to smooth out changes in voltage levels. Relays are remotely controlled electric switches.

43. The correct answer is **(D)**.

    Look up the old part number (390SC) in the *Part Number* column. Read the replacement number (**3988SG**) across in the *New Number* column.

44. The correct answer is **(C)**.

    Both Specialists are correct. A distributorless ignition system has a separate coil for each spark plug or pair of plugs; thus, it has no need for a rotor. On the other hand, some of these systems have the coil mounted directly on the spark plug and do not need spark plug wires.

45. The correct answer is **(D)**.

    The **rectifier** is an electrical device. The muffler, catalytic converter, and resonator are all exhaust system components.

46. The correct answer is **(A)**.

    Dual exhaust pipes may be connected with a **crossover pipe**. Headers are a tubular form of an exhaust manifold, and a resonator is a type of secondary muffler.

47. The correct answer is **(C)**.

    Both Parts Specialists are correct. Lost sales can result from a customer requesting a part the store does not stock or the store stocking brands other than those requested by the customer.

48. The correct answer is **(C)**.

    Both Parts Specialists are correct. Most modern stores track inventory with a computer (see Figure 3-6), replacing the card system of the past.

49. The correct answer is **(D)**.

    A **halfshaft** connects the transaxle to the drive wheels. A transaxle requires a halfshaft for each drive wheel. A countershaft is found in a manual transmission. Line axles and line shafts are not used in automotive applications.

Figure 3-6 Inventory is usually tracked with a computer.

50. The correct answer is **A**.

    The **universal joint (U-joint)** connects the driveshaft to the pinion yoke of the differential. *Pinion joint* is a false term. Halfshafts are used to transfer torque from a transaxle to the drive wheels. The *pinion nut* secures the yoke to the pinion gear.

51. The correct answer is **D**.

    All of the answers are correct. A limited-slip differential uses internal clutch plates to limit the slip and speed difference between the drive wheels. This arrangement prevents wheel spinning and improves traction for acceleration. It requires a special lubricant or an additive to the differential oil.

52. The correct answer is **B**.

    Anti-lock brakes operate by pulsing the pressure to the wheel and caliper cylinders to prevent the lockup of any one wheel. When the anti-lock feature is being activated, **it is normal to feel a pulsation of the brake pedal**. This pulsation should occur only during times of hard braking, and the driver must maintain pedal pressure to keep it activated. By preventing wheel lockup, the anti-lock brakes help maintain traction on a slippery road surface. This action reduces skidding and improves driver control of the vehicle.

53. The correct answer is **(C)**.

    Both Specialists are correct. The illustrated part is a ball joint. It allows the steering knuckle to turn right and left as well as permitting the control arm to move up and down. Worn ball joints will allow changes in front axle alignment angles, causing excessive tire wear.

54. The correct answer is **(B)**.

    Passenger compartment temperature is regulated by using a blend door to **mix the correct ratio of incoming heated air and fresh air**. Fresh air entering the vehicle is split through two passageways. One is directly into the passenger compartment and the other is by way of the heater core. The temperature of the engine coolant is determined by the engine operating temperature. Vehicles prior to the 1980s controlled heated air temperature by varying the volume of coolant passing through the heater core. With an air conditioning system, the compressor will cycle to maintain the lowest possible temperature of the fresh air, and then the heater core and blend door will be used to raise the temperature to the desired level.

55. The correct answer is **(B)**.

    A heavy duty flasher should be used when a standard duty flasher flashes too fast. The speed at which a flasher will flash the turn signal or hazard lights is related to the amount of current flowing to the lights. More lights (e.g., connecting a trailer) will require more current flow and cause the flasher to cycle faster. With the same current load, a standard flasher will cycle faster than a heavy duty flasher.

56. The correct answer is **(D)**.

    Bolt diameter is measured by the **diameter of the threads**.

57. The correct answer is **(B)**.

    Metric thread pitch is the distance between the threads measured in millimeters. American National Standard (English) thread pitch is the number of threads per inch.

58. The correct answer is **(D)**.

   Studs do not have a head and must be installed and removed with two nuts locked together or a special tool. Normally, a stud is tightened into one object and a nut, on the other end of the stud, is used to clamp another object to the first object. *Screws, nuts,* and *bolts* all have some type of head with which to turn them.

59. The correct answer is **(D)**.

   Keep your back as straight as possible when lifting a load—**do not bend your back over the object**. Lift with the legs, not with the back. Keep the object as close to the body as possible to maintain balance. Think ahead; make sure a clear walkway is available for carrying the object once it is lifted.

60. The correct answer is **(C)**.

   Both Specialists are correct. Most of the chemicals found in the parts store are hazardous if ingested or if they come in contact with the body. Information regarding the hazards, safe handling, antidotes, and technical specifications of a chemical are found on its Material Safety Data Sheet (MSDS). Under some circumstances, a copy of the MSDS must accompany the sale of the chemical.

61. The correct answer is **(D)**.

   All of the statements are true regarding the disposal of hazardous waste. The use of a biodegradable solvent or lubricant will make disposal much easier. Any caustic solutions which are used should be neutralized prior to their disposal. Hazardous waste handling can also be simplified by contracting with a hazardous waste disposal company.

62. The correct answer is **(C)**.

   A bolt with three radial lines embossed on the head would be a **grade 5**. The American National Standard bolt grade is two more than the number of radial lines embossed on the head (3 on the head plus 2 equals 5). No markings is grade 2, five markings is grade 7, etc. Metric bolt grade marking is with a decimal number: 5.8, 9.8, 10.9, etc.

63. The correct answer is **(D)**.

Flare, compression, and pipe are the three types of fittings. A *flare fitting* requires that the end of the tubing be expanded at an angle or flared. This operation requires the use of a flaring tool. *Compression fittings* do not require any special tools. A soft metal band, known as a ferrule, is placed between the end of the tubing and the fitting. As the two fittings are tightened together, the ferrule is crimped into a sealing connection between the fittings and the tubing. *Pipe fittings* seal using tapered threads. The more that pipe fittings are tightened, the closer the threads are brought together.

64. The correct answer is **(A)**.

Pipe threads are tapered. The further a pipe is screwed into the fitting, the tighter the threads jam together.

65. The correct answer is **(C)**.

Both Specialists are correct. Information regarding the proper application of a chemical can be found on the product label or any accompanying Material Safety Data Sheet (MSDS). In some instances, the sale must include a copy of the MSDS.

Figure 3-7 Vehicle Identification Number (VIN) location.

66. The correct answer is **D**.

The standard location for the Vehicle Identification Number (VIN) is **attached to the driver-side top of the instrument panel and visible through the windshield** (see Figure 3-7).

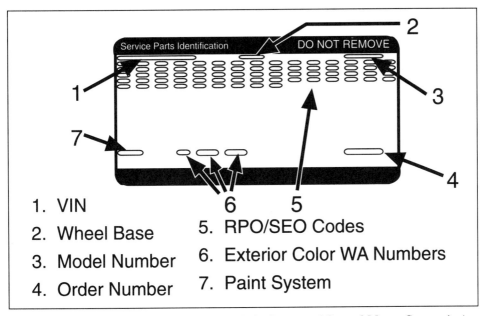

Figure 3-8 Service parts identification label *(Courtesy of General Motors Corporation)*.

67. The correct answer is **A**.

   Paint color is not found in the Vehicle Identification Number (VIN). It is located on a separate trim tag, called the service parts identification label (see Figure 3-8), which is mounted on the body, usually in the radiator area or on the firewall. The other items are part of the VIN.

68. The correct answer is **B**.

   The **inboard Constant Velocity (CV) joint** allows for changes in the angle of the halfshaft and **allows for shaft movement in and out** of the joint as the shaft moves up and down, effectively changing its length. The outboard Constant Velocity (CV) joint allows for changes in the angle of the halfshaft and steering motion of a front wheel but not for thrust angle. Thrust angle refers to the alignment of the rear axle to the vehicle.

69. The correct answer is **B**.

   Part #6 in Figure 2-4, page 28, is the parking brake strut. Its function is to **apply the parking brake.**

70. The correct answer is **A**.

   The customer will NOT need **brake spring pliers**. Extra brake fluid may be needed. While replacing the brake pads, the customer should also rebuild the calipers using a caliper rebuild kit and pack the wheel bearings with wheel bearing grease. Brake spring pliers are needed only when servicing drum brakes.

71. The correct answer is **C**.

   The **fourth digit in the chart** refers to the GVWR Range/Brake System. In this example, code D identifies the 5001-6000 pound range.

Figure 3-9 A core charge applies to parts that are turned in to be remanufactured or rebuilt.

72. The correct answer is **(D)**.

All parts are not subject to a core charge. A core charge applies only to those parts that are turned in to be remanufactured or rebuilt for later resale (see Figure 3-9). It is not charged if the customer brings in the used part at the time of purchase. Core charges are refunded when the customer brings in the used part if it is not worn too badly to be rebuilt. The exchange parts program allows the customer to get good quality parts at a lower cost than that for new parts.

73. The correct answer is **(A)**.

During an engine overhaul, **oversize pistons** must be installed if the cylinders are bored. Undersize rod and main insert bearings are used when the crankshaft journals are turned. Crankshaft journals must be turned after wear has eroded the metal to the point that the journal is no longer smooth, round, or the correct size. This wear can be caused by trash in the motor oil or by a lack of lubrication. Camshaft journals are not usually machined since normal wear of the lifter lobes requires that the camshaft be replaced.

74. The correct answer is **(D)**.

The thermostat **controls the water flow through the radiator**, preventing water flow through the radiator until the engine is at operating temperature. A normally operating thermostat sets the minimum engine operating temperature and will have no effect on the maximum temperature or the air flow through the radiator. A good thermostat decreases engine warm-up time.

75. The correct answer is **(D)**.

All of the answers are correct. The function of the engine fan is to increase air flow through the radiator during periods of low vehicle speed. This is most important while driving in slow traffic or when waiting at a stop light. Engine fans may be mounted on the end of the water pump and driven by a belt or mounted on the radiator and driven by an electric motor.

76. The correct answer is **(B)**.

   Parts Specialist B is right. Additional forms must usually be completed for warranty-returned parts. This allows the manufacturer to determine what is wrong with the part and to credit the store for the return. The part should never be placed back in the inventory to be sold to another customer.

77. The correct answer is **(A)**.

   Each refers to a single item. *Carton* or *case* refers to a number of parts packaged together and sold as a unit from the supplier. *Pair* refers to two of something; usually both are needed to complete the job.

78. The correct answer is **(C)**.

   The first two digits indicate **1985.** In addition to the numbers shown in Figure 2-6, page 31, some codes also include the production Julian date, which is the day number of the year, January 1 being 001 and December 31 being 365. (Example: a Julian date of 002 would mean the transmission was produced on the second day of the year—January 2.)

79. The correct answer is **(A)**.

   The **fuel pump** moves the fuel from the tank to the carburetor. A fuel injector sprays fuel into the intake manifold on vehicles not equipped with a carburetor. A fuel regulator may be used to control the pressure developed by the fuel pump. A fuel rectifier is not an actual part.

80. The correct answer is **(A)**.

   The oxygen sensor is located **between the exhaust manifold and the catalytic converter** or in the catalytic converter. Its function is to monitor the oxygen content in the exhaust gas. The signal from the oxygen sensor tells the computer whether a rich or lean condition exists in the air/fuel ratio of the engine. The computer can vary the amount of fuel entering the engine to achieve the optimum ratio. Oxygen sensors are used on computer-equipped vehicles having either carburetor or fuel injection systems.

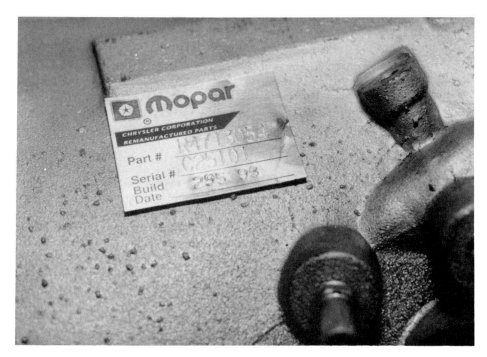

Figure 3-10 All major automobile components have an ID number stamped on them.

81. The correct answer is **A**.

    Component ID data is **stamped on all major components of the automobile** (see Figure 3-10). It is very valuable in identifying the parts needed to make the necessary repairs.

82. The correct answer is **D**.

    Neither Specialist is correct. In reference to the chart, an RPO code of GU6 identifies the rear axle (3.42) ratio, and a code of K60 identifies the generator (100 amp).

83. The correct answer is **(C)**.

   Both Parts Specialists are correct. Hatchbacks and fastbacks are similar body styles with sloping rear glasses over the trunk area. On a *hatchback*, the rear glass will lift to allow access to the trunk. The *fastback* has a fixed rear glass; access to the luggage compartment is from the inside of the vehicle.

84. The correct answer is **(D)**.

   The **Positive Crankcase Ventilation (PCV)** system draws hydrocarbons from the engine crankcase and routes them through the intake manifold to be burned in the engine. *Air Injection Reaction (AIR)* pumps air into the exhaust system to burn hydrocarbons, carbon monoxide, and oxides of nitrogen coming from the engine. *Exhaust Gas Recirculation (EGR)* meters some exhaust gas into the intake manifold to reduce the combustion chamber temperature and prevent the formation of oxides of nitrogen. *Early Fuel Evaporization (EFE)* is used to warm the air going into a cold engine to improve driveability and reduce hydrocarbon and carbon monoxide emissions.

85. The correct answer is **(A)**.

   Carbon dioxide, which is exhaled by animals, is necessary for plant life. Carbon dioxide and water vapor are produced in the catalytic converter by combining oxygen, hydrocarbons, and carbon monoxide. Carbon monoxide, hydrocarbons, and oxides of nitrogen are the main pollutants from a gasoline engine.

86. The correct answer is **(B)**.

   Exhaust Gas Recirculation (EGR) reduces oxides of nitrogen by metering some exhaust gas into the intake manifold to reduce the combustion chamber temperature. If this temperature is below 2,500°F, oxides of nitrogen will not be produced. The other answers reduce pollution, but not specifically oxides of nitrogen. *Air Injection Reaction (AIR)* pumps air into the exhaust system to burn hydrocarbons, carbon monoxide, and oxides of nitrogen coming from the engine. *Early Fuel Evaporization (EFE)* is used to warm the air going into a cold engine to improve driveability and reduce hydrocarbon and carbon monoxide emissions. *Positive Crankcase Ventilation (PVC)* draws hydrocarbons from the engine crankcase and routes them through the intake manifold to be burned in the engine.

Figure 3-11 Most service information is computer accessible.

87. The correct answer is **(D)**.

    All of these are sources of information which can help the Specialist meet the needs of the customer. *Service manuals* contain the diagnosis and repair procedures for the vehicle systems. Also included are the various specifications needed for repair and adjustments. *Service bulletins* are supplied by the vehicle manufacturers and parts suppliers. They provide supplemental and update information for the service manuals. Both service manuals and service bulletins often are supplied on computer disks (see Figure 3-11). *Supplier training materials* are used to train the Parts Specialist in new automotive systems and their related repair procedures.

88. The correct answer is **(B)**.

    The paint code is on the **trim tag** which is usually located on the firewall, on the body near the radiator, or in a door post area.

89. The correct answer is **B**.

   The pilot bearing is **located in the center of the flywheel**. This bearing supports the end of the transmission input shaft, not the clutch-release lever. The throw-out bearing, also known as the clutch-release bearing, is mounted on the clutch-release lever.

90. The correct answer is **D**.

   All of these devices are used. In order to engage the gear ranges of an automatic transmission, various parts of the gearset must be held by bands or be turned or held by clutches. *Servos* are the hydraulic pistons which apply the bands.

91. The correct answer is **C**.

   Both Specialists are correct. The torque converter slips and allows the engine crankshaft and automatic transmission input shaft to rotate at different speeds. This has the same effect as releasing the clutch on a manual transmission. During acceleration, the torque converter provides torque multiplication, a hydraulic form of gear reduction which places a 2:1 gear ratio between the engine and the transmission.

92. The correct answer is **D**.

   The customer should pay **$99.97**, the amount of the Total Invoice. The $6.55 amount is the tax on the purchase. The $66.95 amount is the selling price of one camshaft. The $93.42 amount is the total of the parts purchased without the sales tax.

93. The correct answer is **A**.

   Look up the number AB1220 in the *Repl. No.* column. Verify that it is manufactured by ABCD by referencing the adjacent *Mfg.* column. The corresponding part number from the inventory (**15244**) will be listed on the same line under *Part No.*

94. The correct answer is **D**.

   The maximum battery voltage is controlled by the **voltage regulator**. It may be inside the alternator or mounted externally. Rectifiers and diodes are used in the alternator to convert AC to DC. If they are defective, the battery will not be charged. The ignition coil is part of the ignition system and does not affect battery charging operation.

95. The answer is **C**.

   Both Specialists are correct. Either a defective alternator or voltage regulator can cause a battery not to be charged. A defective alternator will not produce the current needed to charge the battery. The voltage regulator sets the maximum charging voltage produced by the alternator. If a defective voltage regulator sets the charging voltage at too low a level, the battery will not be charged.

Figure 3-12 The parts store must display both SAE and metric nuts and bolts.

96. The correct answer is **D**.

   Neither Specialist is correct. Metric threads are becoming the worldwide standard. American built cars are being converted to metric threads as production changes allow. Components designed prior to the mid-1970s use English threads, while newer components are using metric threads. The parts store must display both types (see Figure 3-12).

97. The correct answer is **C**.

   Look up the two part numbers with an I.D. of 5.036 (399-1563 and 399-1561). The correct spring will be the one with a wire diameter of 0.563 (**399-1561**).

98. The correct answer is **C**.

Look up *Chr* in the left column of the abbreviation chart. The corresponding name (**Chrome**) is read across the chart to the right.

99. The correct answer is **C**.

Page **353** will have tapered roller bearing illustrations. Page 300 will list bearing interchanges. Page 352 will give general information on bearing sets with competitor interchange. Page 435 will present cylindrical bearing descriptions and illustrations.

100. The correct answer is **D**.

Neither Parts Specialist is right. Supersession bulletins note part numbers that now supersede (replace) previous part numbers. These are not correction bulletins, which correct catalog errors due to typing errors or inaccurately assigned part numbers. Technical bulletins explain unusual installation, fit, or maintenance problems.

101. The correct answer is **A**.

The cable length should be adjusted if the parking brake will not hold. The parking brake is mechanical and the brake fluid level (hydraulics) will have no effect on its operation.

102. The correct answer is **D**.

The **bushings** connect the control arm to the frame. The *ball joint* connects the spindle to the control arm. The *coil spring* supports the weight of the vehicle. The *idler arm* is a part of the steering system, connecting the center link to the frame.

103. The correct answer is **A**.

The **tie-rod end** is used on all steering systems. The drag link, idler arm, and pitman arm are used on a parallelogram steering system, not a rack-and-pinion.

104. The correct answer is **C**.

The **recirculating ball worm gear and rack-and-pinion** are the two types of steering gearboxes. The MacPhearson strut is part of a suspension system. The pitman arm is used with the recirculating ball worm gear steering gearbox and connects to the center link.

Figure 3-13 Every store item must be counted during the inventory process.

105. The correct answer is **A**.

    This sheet would be used to perform a **physical inventory**. During the inventory process, each part is manually counted and the number on hand is written on the form (see Figure 3-13). Usually, a second count is done by someone else to verify the amount. Once the on-hand quantity is determined, it is entered into the computer for each part number. The part numbers on this sheet are arranged by the bin location to make the job of counting easier. An *invoice* is a record of the sale to a customer. An invoice sheet identifies the customer, the quantity of the parts sold, and the sale price. A *stock order* is used to order inventory from the suppliers. It would include quantities for the amount of merchandise being ordered. *Cataloging* is the process of looking up the needed parts in the parts catalog. This normally has similar parts grouped together according to year, engine size, vehicle model, etc.

106. The correct answer is **D**.

    Neither Parts Specialist is right. Refer to the *Locator-Pad* to look up the part numbers. Read down the *Part-No.* column to 355798. The column labeled *On-Hand* is the number of the part in stock. There are five parts with the number 355798 in stock, not two. Find part number 332395 and read across to see that it is located in bin 785 instead of bin 741.

107. The correct answer is **A**.

To **ignore the problem** will not solve it. Verify the discrepancy by making a note of the suspected part numbers. Verify the bin count to confirm the problem, and then notify the supervisor.

108. The correct answer is **B**.

Stock rotation is **selling the older stock before the newer**. This prevents the older stock from sitting at the back of the shelf and eventually having an aged appearance. Rotating the product so the labels are toward the front of the shelf is known as *facing*. Rearranging the shelves for a neater appearance is done whenever necessary to increase customer appeal.

109. The correct answer is **C**.

Both Specialists are correct. A special order is placed whenever a customer purchases an item not kept in stock. A part which repeatedly appears on Lost Sale or Special Order reports would be considered for stocking status.

110. The correct answer is **C**.

Both Specialists are correct. A port fuel injection system has a fuel injector for each cylinder positioned in the intake manifold at the base of the intake valve. Fuel is sprayed directly into the cylinder. A throttle body injection system has one or two injectors mounted in a throttle body at the inlet end of the intake manifold. Fuel is delivered to the cylinders through the intake manifold in the same way as a carburetor delivers fuel.

111. The correct answer is **B**.

The heat range of a spark plug indicates **how quickly the plug transfers heat from the plug tip to the cylinder head**. If the heat is transferred slower, the plug is said to be hotter. The heat range is indicated by the number assigned to the spark plug. The required heat range is determined by engine design characteristics, not by climate, coil size, plug life, and driving conditions.

112. The correct answer is **D**.

The electricity needed to create the spark for combustion is produced by the **ignition coil**. The distributor and rotor route the voltage to the correct spark plug. The alternator recharges the battery.

113. The correct answer is **B**.

   In order to maintain an adequate supply of used parts for remanufacturing, a core charge is added when the customer buys a **remanufactured part**. This charge is refunded when the customer returns the used part from the vehicle. A freight charge is added to special order parts to cover their transportation to the store. Some parts may have a restocking fee that is charged if they are returned for a refund.

114. The correct answer is **C**.

   Both Specialists are correct. A **set** is usually the number of a part needed to complete the job. If only a partial set is needed, usually a **special order** must be placed for the lesser quantity.

115. The correct answer is **C**.

   Replacement parts should be ordered for the broken kit so it can be restocked and sold as new once complete. It should not be sold as a complete unit until the missing parts are replaced. If replacement parts are not available, it could be sold for spare parts at a reduced price.

116. The correct answer is **A**.

   Main bearing caps secure the crankshaft to the block. Rod bearing caps hold the rod to the crankshaft. The other parts are not related to the crankshaft.

117. The correct answer is **D**.

   All three methods are used to drive a camshaft. The overhead valve arrangement can use either a timing chain or gears. The overhead cam system uses either a timing belt or chain.

118. The correct answer is **A**.

   The **timing belt** on an overhead cam engine is normally replaced whenever the head is removed. The lifters are replaced only during an engine overhaul or when they are defective. Pushrods are not used on overhead cam engines.

119. The correct answer is **D**.

   Neither Parts Specialist is correct. Either an exchanged or returned part will affect the on-hand inventory if an adjustment is not made. With an *exchanged* part, the inventory will indicate too few of the returned part and too many of the part which was exchanged. With a *returned* part, the inventory count will show a shortage of the part number.

Figure 3-14   Items displayed at eye level are easily noticed.

120. The correct answer is **(B)**.

    Customers are most likely to notice items placed at **eye to chest level** (see Figure 3-14). They are less likely to find items on the top and bottom shelves or stacked on the floor.

121. The correct answer is **(D)**.

    Placing high-volume sales items **along the walls and away from the entrance** will increase customer traffic throughout the entire store. This arrangement helps increase impulse sales. Concentrating these items together in such areas as close to the cash register or the door or in the center of the store will decrease customer traffic and impulse purchases.

Figure 3-15 Items visibly displayed are usually individually priced.

122. The correct answer is **(B)**.

    All **items displayed in front of the parts counter** should be individually priced (see Figure 3-15). This placement encourages customers to shop and increases impulse sales. Parts in the stock room are not usually individually priced since the counterperson must retrieve them before selling them to the customer.

123. The correct answer is **(C)**.

    Both Specialists are correct. Both situations involve customer satisfaction, which is a top priority in the sales business. The Specialist should get whatever help is necessary in order to sell the correct parts and satisfy the customer. In addition, experienced employees are often asked to help train the new employees.

124. The correct answer is **D**.

It is the responsibility of all employees to **keep all areas of the store** clean and orderly. All employees suffer from lost sales if the floors, shelves, and displays are a safety hazard or do not appeal to the customer.

125. The correct answer is **A**.

Parts Specialist A is correct—a battery is dangerous. The sulfuric acid used in an automotive battery is very corrosive. It can cause severe burns if it comes in contact with skin or clothing. Use care when handling batteries. In addition, be careful when charging a battery. It gives off hydrogen gas, which is very explosive. Sparks and open flames should be avoided around batteries being charged.

126. The correct answer is **A**.

By **answering the phone and asking the caller to hold**, the Specialist will be able to finish serving the walk-in customer, and then take care of the caller. This allows the customers to be handled in a fair and courteous manner. Neither the phone call nor walk-in customer should be ignored. The Specialist will not be able to fairly handle both customers at the same time.

127. The correct answer is **D**.

The parts **counterperson** is most likely to sell machine work to the customer. The counterperson is in direct contact with the customer when selling the related parts that accompany the machine work. Promoting machine work also usually means more parts sold across the parts counter.

128. The correct answer is **D**.

The **larger number** indicates a finer grit.

129. The correct answer is **B**.

Vacuum hose is normally used on the windshield washer system. Fuel hose is reinforced and made of a chemical-resistant material. Vacuum hose is not reinforced and will deteriorate when in contact with gasoline and other hydrocarbons. Fuel hose should be used for transferring gasoline through the fuel system and fuel vapors through the emissions system.

Figure 3-16 A single display can usually handle rubber hose and tubing.

130. The correct answer is Ⓒ.

    Both Specialists are correct. Copper and steel tubing is measured by the outside diameter. Vacuum hose is measured by the inside diameter. A single display can be used to dispense several types of hose and tubing in an attractive and efficient manner (see Figure 3-16).

GLOSSARY

Accumulator — Part of an air conditioning system which contains a desiccant that absorbs moisture from the refrigerant.

Air filter — Part of an engine intake system which cleans dirt and dust from the air before it enters the engine.

Air Injection Reaction (AIR) — An emissions control system which pumps air into the exhaust system to burn hydrocarbons, carbon monoxide, and oxides of nitrogen coming from the engine.

Air pump — Part of the Air Injection Reaction (AIR) system which forces extra air to mix with the exhaust gases. This action causes the continued burning of any hydrocarbons and carbon monoxide remaining in the exhaust.

Alphanumeric — A numbering system consisting of a combination of letters and numbers. They are placed in order starting from the left digit and working across to the right. This system is commonly used in parts catalogs and price sheets.

Alternating current (AC) — An electrical current which flows alternately in two directions, forward and backward. It is produced by some form of mechanical device or motion, such as an alternator. Alternating current cannot be used to charge a battery. It must first be converted (rectified) to direct current for battery charging.

Alternator — The electrical device driven by the engine which recharges the battery. It produces alternating current using rotating field coils inside stationary stator windings.

Anti-lock brakes — A brake system which operates by pulsing the pressure to the wheel and caliper cylinders to prevent the lockup of any one wheel, thus preventing a skid.

Automatic transmission — A transmission which automatically selects the correct gear ratio needed for the driving conditions. The driver has only to select the direction of travel desired.

Back order — Merchandise ordered from a supplier but not shipped due to the supplier being out of stock.

Ball joint — Part of the steering system which connects the spindle to the control arm. It allows the steering knuckle to turn right and left as well as permitting the control arm to move up and down.

Band — A flexible flat piece inside an automatic transmission which holds parts of the gearset to create the correct gear ratio.

Battery — A device within the electrical system which stores voltage until it is needed for vehicle operation.

Blend door — Part of the air conditioning/ventilation system which controls the ratio of incoming heated air and fresh air. This ratio is adjusted to control the temperature of the passenger compartment.

Blowby — The unburned fuel and combustion byproducts which leak past the piston rings and into the crankcase.

Brake pads — The friction elements of a caliper-type brake system. They are usually forced toward the rotor by hydraulic pressure.

Brake shoes — The friction elements of a drum-type brake system. They are usually expanded outward by hydraulic pressure to contact the inside diameter of the brake drum.

Brake spring pliers — Pliers used in the disassembly of brake drum components.

Bushings — A smooth cylinder used to reduce friction and to guide the motion of the parts.

Camshaft — The shaft in an engine which causes the valves to open and close at the correct times.

Capacitor — An electrical device used to smooth out changes in voltage levels.

Carbon dioxide — A colorless, odorless, incombustible gas which is exhaled by humans and required by plant life. In the automotive emissions system, carbon monoxide is mixed with air (oxygen) to form harmless carbon dioxide.

Carbon monoxide — A colorless, odorless, toxic gas formed by internal combustion engines.

Carburetor — A device used on some engines to mix the fuel and air in the correct ratio for efficient combustion.

Cataloging — The process of looking up the needed parts in the parts catalog.

Catalytic converter — A metal canister mounted in the exhaust system containing metals that convert harmful exhaust gases into safer gases. A catalytic converter speeds up the reaction but is not consumed in the chemical reaction.

Caustic — A compound that is able to burn, corrode, or eat away another compound.

Celsius — The metric unit of temperature.

Check valve — A valve which allows something to move or flow in only one direction.

Circuit breaker — An electrical device which protects the circuit by interrupting the current flow when it exceeds the rated capacity. They may be reset manually or automatically.

Closed-loop — An operating state in a computer-controlled engine in which the computer is controlling engine operation based upon information created by the sensors.

Closing a sale — Getting a commitment to buy from a customer.

Clutch — The part of a driveline system which is used to interrupt the power flow between the engine and the transmission.

Clutch-release lever — The part of a manual transmission system which operates the clutch-release (throw-out) bearing.

Coil spring — A spring which is wound into a spiral shape. Coil springs are commonly used on automotive suspension systems.

Combustion chamber — The area inside the cylinder head and block where the burning of the fuel takes place.

Compression — The act of forcing something together. In an automotive engine, the air and fuel are compressed in the cylinders to create a larger explosion and thus more power upon ignition.

Compressor — The part of the air conditioning system which compresses the refrigerant vapor and pumps the refrigerant.

Condenser — The part of the air conditioning system which cools the hot vapor and converts it to a liquid. The condenser is usually mounted in front or on top of a vehicle for better air flow.

Constant Velocity (CV) joint — Part of the drivetrain which allows for changes in the angle of a driveshaft or halfshaft.

Control arm — Suspension parts which control spring action and the direction of travel of the axle as it reacts to driving conditions.

Core charge — A charge which is added when the customer buys a remanufactured part. Core charges are refunded to the customer when he/she returns a rebuildable part.

Correction bulletin — A bulletin which corrects catalog errors due to printing errors or inaccurately assigned part numbers.

Countershaft — A shaft used in transmissions to transfer the motion from the input shaft to the output shaft.

Crankcase — The lower half of an engine.

Crankshaft — The shaft in an engine which converts the reciprocating piston motion into rotary motion for the driven device.

Crossover pipe — The pipe which connects the two exhaust pipes of a dual exhaust system.

Cubic inch — The volume equal to a cube with one-inch sides. The term is commonly used to describe engine displacement.

Customer relations — A description of how a salesperson interacts with the customer.

Cylinder head — The removable covering mounted on top of the cylinders. It seals the combustion chamber and usually contains the valves and spark plugs.

Desiccant — Any substance used to absorb moisture. Desiccant is used in an air conditioning system to keep moisture from forming corrosive compounds.

Diameter — The distance straight across a circular figure; the largest measurement which can be taken across a circular object.

Differential — The set of gears which transmits power from the driveshaft to the wheels and allows the drive wheels to turn at different speeds for cornering.

Direct current (DC) — An electrical current which flows in only one direction. It is usually created from a chemical source, such as a battery, and can be used to recharge a battery.

Discount — The amount of savings being offered to a customer, usually expressed as a percent.

Distributor — The part of the ignition system which directs the secondary voltage (spark) to the correct spark plug.

Do-it-yourselfer — Someone who performs service and repair on his/her own vehicle.

Drag link — The rod which connects the steering box to the steering knuckle on a straight axle front.

Driveshaft — The shaft which connects the transmission to the differential.

Dual exhaust — An exhaust system which uses separate exhaust pipes for each bank of cylinders.

Early Fuel Evaporization (EFE) — An emissions system used to warm the air going into a cold engine to improve driveability and reduce hydrocarbon and carbon monoxide emissions.

Emergency order — An order placed out of the normal cycle of stock orders, usually when unexpected or emergency parts are needed.

Engine block — The main body of an engine. The block contains the cylinders and carries the accessories.

Engine coolant — The solution used in an engine to carry heat away from the cylinders. It is usually a 50/50 mixture of ethylene glycol and water.

Evaporator — The portion of an air conditioning system which is mounted near the passenger compartment. It contains low-pressure refrigerant gas and removes heat from the air forced around it.

Exhaust Gas Recirculation (EGR) — An emissions device which meters some exhaust gas into the intake manifold to reduce the combustion chamber temperature and to prevent the formation of oxides of nitrogen.

Exhaust manifold — The part of the exhaust system which bolts directly to the cylinder heads. Exhaust gases leaving the cylinder are routed through the cylinder head to the exhaust manifold.

Facing — Rotating the product so the labels are toward the front of the shelf. Facing improves shelf appearance and customer appeal.

Fahrenheit — The temperature unit used in the English system of measurement.

Fastback — A vehicle body style which has a fixed rear glass. Access to the luggage compartment is from the inside of the vehicle.

Flare — A type of tubing connection which requires that the end of the tubing be flared or expanded at an angle.

Flasher — The part of the lighting electrical system which causes the turn signals and hazard lights to blink.

Freight charge — A charge added to special order parts to cover their transportation to the store.

Fuel injector — The fuel system component which sprays fuel into the intake manifold on vehicles not equipped with a carburetor.

Fuel pump — The fuel system component which moves the fuel from the tank to the carburetor or the fuel injection unit.

Fuse — An electrical device which protects the circuit by interrupting the current flow when the flow exceeds the fuse's rated capacity. Fuses are constructed of a conductor encased in plastic or glass and must be replaced when blown.

Fusible link — An electrical device which protects the circuit by interrupting the current flow when the flow exceeds the fusible link's rated capacity. A fusible link is constructed of a wire surrounded by a special insulation and must be replaced when blown.

Grade — A method of classifying bolt strength. The grade of a bolt is indicated by markings on the head.

Halfshaft — One of two shafts which connect a transaxle to the drive wheels.

Hatchback — A vehicle body style in which the rear glass will lift to allow access to the trunk.

Headers — A tubular form of exhaust manifold which is used to increase exhaust gas flow and improve performance.

High-volume — Describes a popular item which is sold in large numbers.

Hydrocarbon — An organic compound made up of hydrogen and carbon. Most automotive fuels and lubricants contain hydrocarbons, a common source of pollution. As an automotive term, it refers to unburned fuel in the exhaust.

Hydrogen — A colorless, highly flammable gas which is the most common in the universe. It is used in the production of methanol, an automotive fuel and fuel additive.

Idle — A condition in which the engine is running at a low speed.

Idler arm — A part of the steering system which connects the center link to the frame.

Ignition coil — A part of the ignition system which produces the electricity needed to create the spark for the spark plugs.

Ignition system — The part of the automotive electrical system which creates and delivers the high-voltage spark to the spark plugs.

Impulse item — Something the customer did not intend to buy when entering the store.

Inboard — A position reference relating to being toward the center of the vehicle.

Individually priced — The condition of having each part on the display priced for customer convenience.

Input shaft — The shaft which carries torque into the transmission.

Insert bearings — A bearing surface consisting of a split circular shell which is inserted between the engine block or connecting rods and the crankshaft.

Intake manifold — The part of the intake system which bolts directly to the cylinder heads. The air or air/fuel mixture must pass through the intake manifold before entering the cylinder head and combustion chamber.

Intake valve — The valve in the intake port which opens to allow the air/fuel mixture to enter the combustion chamber.

Intercooler — A part of the intake system used with a turbocharger to cool the air entering the intake manifold.

Inventory — The goods a store has in its possession for resale.

Invoice — The record of a sale to a customer.

Journal — The part of a shaft which is supported by and comes in contact with a bearing.

Julian — A calendar system devised by Julius Caesar which numbers the days consecutively starting with January 1. The Julian calendar is often used in production codes and computer systems.

Lifter — A component of the automotive valve train which converts the rotary motion of a camshaft lobe into the reciprocating motion needed to open and close the valves.

Limited-slip differential — A differential which uses internal clutch plates to limit the slip and speed difference between the drive wheels. This limiting improves traction on slick surfaces and helps eliminate wheel spinning.

Liter — A unit of volume used in the metric system of measurement. A liter is slightly larger than a quart in the English system.

Locator-pad — A quick-reference chart which will list a parts description, on-hand quantity, cost, and location in part number order.

Lost sale — A customer not purchasing a requested item either because the store does not stock the item or because the store stocks brands other than the one requested by the customer.

Machine work — Work done in a machine shop. In automotive applications, such work refers to engine block boring, shaft journal turning, brake drum and rotor turning, and cylinder head repair services.

MacPhearson strut — A suspension system component which combines a lower lateral link with a vertical strut to combine the features of a spindle and a shock absorber.

Main bearing caps — The caps which secure the crankshaft to the block.

Material Safety Data Sheet (MSDS) — A sheet which accompanies chemicals and contains information regarding the proper application along with the safety and environmental concerns.

Micrometer — A precision measuring tool which can measure to dimensions of 0.0001 inch or centimeter.

Muffler — The part of the exhaust system which is used to reduce exhaust noise.

No-return policy — A store policy that certain parts cannot be returned after purchase. It is very common on electrical and electronic parts.

Oil filter — A device on the engine for removing dirt, carbon, and other impurities from the lubricating oil.

On-hand — The quantity of an item that the store has in possession.

Open-loop — An operating state in a computer-controlled engine in which the computer is controlling engine operation based upon a pre-determined program. It is usually in effect until the engine sensors signal that the engine has reached operating temperature.

Orifice tube — A part of an air conditioning system which causes the drop in refrigerant pressure that results in the change in temperature necessary for cooling.

Outboard — A position reference indicating being toward the outside of the vehicle.

Overhead cam — An engine valve train system which has the camshaft positioned on top of the cylinder head.

Oxides of nitrogen — Pollutants formed when nitrogen and oxygen come in contact with combustion chamber temperatures in excess of 2,500 degrees Fahrenheit.

Oxygen — A gas making up approximately 18 percent of the air in the atmosphere and required for the combustion process to take place in an engine.

Oxygen sensor — A computer system sensor which monitors the oxygen content in the exhaust gas. The signal from the oxygen sensor tells the computer whether a rich or lean condition exists in the air/fuel ratio entering the engine.

Parking brake — A mechanical brake on the vehicle used for parking or emergency stopping situations.

Parts Specialist — Someone who is employed to sell parts at a parts counter.

Physical inventory — The process whereby each part is manually counted and the number on hand is written on a form or entered into a computer.

Pilot bearing — A bearing mounted in the center of the crankshaft which supports the end of the transmission input shaft.

Pinion yoke — The yoke mounted on the end of the pinion gear of the differential. The pinion yoke transfers torque from the driveshaft to the pinion gear.

Port fuel injection — An automotive fuel delivery system which has a fuel injector for each cylinder positioned in the intake manifold at the base of the intake valve.

Positive Crankcase Ventilation (PCV) — An emissions system which draws hydrocarbons from the engine crankcase and routes them through the intake manifold to be burned in the engine.

Professional — Someone who performs a service as a means of employment.

Profit — The amount received for goods or services above the amount of expenses.

Purchase order — A form giving someone the authority to purchase goods or services for a company.

Pushrod — A valve train component used in engines to connect the lifter to the rocker arm.

R-12 — The trade name for a refrigerant commonly used in automotive air conditioning systems. It is believed to have caused deterioration of the earth's ozone layer. R-12 is being replaced by R-134a as an automotive refrigerant, and its production is being phased out.

R-134a — The trade name for a refrigerant being used in automotive air conditioning systems. It is replacing R-12.

Rack-and-pinion — A steering system which uses a horizontal rack with gear teeth and a pinion gear attached to the end of a steering shaft.

Radiator — The device mounted at the front of the vehicle which is used to cool the engine. The hot coolant flows through the radiator. The radiator fins contain the coolant, and air flowing past the fins will remove heat.

Recirculating ball — A steering system which uses a worm gear filled with ball bearings to reduce steering effort.

Rectifier — An electrical device which converts alternating current (AC) into direct current (DC). The AC current produced by the alternator must be converted to DC in order to charge the battery.

Refrigerant — A substance used to carry heat away from an object. The term usually refers to the chemical used in air conditioning and refrigeration systems.

Relay — An electrical device which allows the remote control of a switch. It normally allows the control of a large current with a much smaller current.

Remanufactured part — A part which has been reconditioned to original standards.

Resonator — A type of secondary muffler.

Restocking fee — The fee charged by the store or supplier for having to handle a returned part.

Return policy — The policy established by the store regarding the return of unwanted or unneeded parts. Return policies may include restocking fees or prohibit the return of certain types of parts, such as electrical or electronic components.

Rod bearing caps — Rod bearing caps hold the rod to the crankshaft.

Rotor — A part of the brake system which turns with the wheel spindle and is clamped by the brake pads for braking action. It is a term which also refers to the rotating part of an alternator that contains the field windings.

Seasonal item — An item which appeals to the customer during only a part of the year. Ice scrapers and lawn equipment are examples of seasonal items.

Service bulletin — A bulletin which provides supplemental and update information for service manuals.

Service manual — A manual which contains the diagnosis and repair procedures for vehicle systems.

Servo — The hydraulic piston which applies the bands found in an automatic transmission.

Shoplifting — The act of stealing goods or display items from a store.

Solvent — A chemical which is able to dissolve another substance. Solvents are normally used in the automotive industry as cleaning agents.

Spark plug — The device which ignites the air/fuel mixture in the combustion chamber.

Special order — An order placed whenever a customer purchases an item not normally kept in stock.

Stock order — A process by which the store orders more stock from the suppliers.

Stock rotation — Selling the older stock on hand before selling the newer stock.

Stud — A fastening device which resembles a bolt. It has threads at each end and no head.

Sulfuric acid — The acid used as an electrolyte in automotive batteries. It is corrosive and produces explosive hydrogen gas while being charged.

Supersession bulletin — A bulletin sent by the parts supplier which lists part numbers that now supersede (replace) previous part numbers.

Technical bulletin — A bulletin which explains unusual installation, fit, or maintenance problems associated with a part.

Thermostat — The device which controls the water flow through the radiator. It prevents water from flowing through the radiator until the engine is at operating temperature.

Throttle body — A fuel injection system which places the fuel injectors in a housing at the location previously used by the carburetor.

Throw-out bearing — A part of the drivetrain used with manual transmissions. It presses in the pressure plate fingers to release the clutch. It is also known as the clutch-release bearing and is mounted on the clutch-release lever.

Thrust angle — The alignment of the rear axle to the vehicle.

Tie-rod end — The movable joint which connects the tie rods to the steering knuckle. Wear in the tie-rod ends will cause excessive tire wear.

Timing belt — A belt which connects the camshaft and crankshaft, synchronizing their rotation. Timing belts are normally used with overhead cam engines.

Timing chain — A chain which connects the camshaft and the crankshaft, synchronizing their rotation.

Timing gears — The gears mounted on the end of the camshaft and crankshaft. They mesh together and synchronize the rotation of the camshaft and crankshaft.

Torque converter — The device located between the engine and the automatic transmission. It will slip and allow the engine crankshaft and automatic transmission input shaft to rotate at different speeds.

Torque multiplication — The process of increasing the torque output of a torque converter. Torque converters usually provide some amount of torque multiplication for vehicle acceleration.

Transaxle — A combination transmission and differential.

Transfer case — A gearbox which connects the front and rear drivelines of a four-wheel drive vehicle.

Transmission — A set of gears which can change ratios to meet the various driving needs of the vehicle.

Transverse — An orientation which refers to the engine and transaxle being mounted crosswise in the vehicle.

Trim tag — A tag located on the vehicle body which outlines the color, body style, trim, and body accessories found on the vehicle.

Turbocharger — A device mounted on the engine which forces air into the intake manifold. Turbochargers are driven by exhaust gas from the engine and serve to boost power and performance.

Universal joint (U-joint) — A flexible coupling which connects the driveshaft to the pinion yoke of the differential.

Vehicle Identification Number (VIN) — A unique number assigned to a vehicle for identification purposes. The VIN number can be decoded for information regarding year of manufacture, manufacturer, body style, engine size, carrying capacity, and other information.

Voltage regulator — The charging system device which sets the maximum charging voltage produced by the alternator.

Warranty return — A defective part returned to the supplier due to failure during its warranty period.

Water pump — The pump located at the front of the engine which circulates engine coolant throughout the cooling system.

Wheel bearings — The bearings located between the wheel hubs and spindle or axle housing and the axle.

Worm gear — A gear cut in such a way that it spirals around a shaft.

APPENDIX

HOW TO REGISTER

ASE tests are given twice a year, in May and November. To obtain a registration application, or free test preparation material, send a card or letter to:

National Institute for Automotive Service Excellence
13505 Dulles Technology Drive, Suite #2
Herndon, Virginia 22071-3421

You can also call ASE at (703) 713-3800; or you can fax them at (703) 713-0727.

NOTES

NOTES

NOTES

NOTES

NOTES

NOTES

NOTES